Contents

Typeset by Jonathan Downes,
Cover and Layout by SPiderKaT for CFZ Communications
Using Microsoft Word 2000, Microsoft Publisher 2000, Adobe Photoshop CS.
First published in Great Britain by CFZ Press

CFZ Press, Myrtle Cottage, Woolsery, Bideford, North Devon, EX39 5QR

© CFZ MMXIII

ISBN: 978-1-909488-09-0

Faculty of the Centre for Fortean Zoology

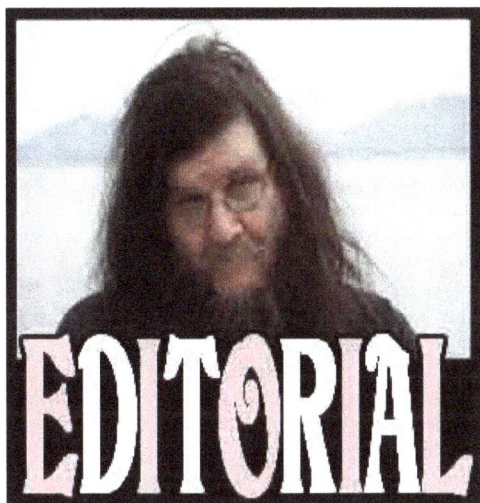

EDITORIAL

Dear friends,

Yes, we are back! And this time it's permanent. I think that probably the first thing that I should do is to explain why there has been a gap of nearly two years since the last issue. There are several reasons - mostly financial.

I dislike the way that in this brave new world of ours even economic disasters have brand names, but we became victims of what has vulgarly become known as 'The Credit Crunch'. Due to a number of factors, but mostly due to the downturn in the global economy, we became perilously close to bankruptcy, and we had to make some severe cutbacks, or go under. One of the cutbacks was the costs associated with the Weird Weekend. Any of you who came to last year's event will have seen that - although it was a very successful conference, and I believe that everyone involved had a jolly good time, certain things that we had done before were not done. For example, there were no lasers; basically the smoke machine has packed up - we are a serious research organisation, and we cannot really justify spending money on something that we use for five minutes once a year. It was fun whilst it lasted, but not in these uncomfortable days of global recession.

And, unfortunately, mainly because before I was reigned in, whenever I have my editorial hat on, I am not very good at counting the pennies, each issue of *Animals & Men* was costing too much to produce, and most importantly, too much to mail out, especially in the UK.

I also thought that it was time for a change. *Animals & Men* had been continuing with practically the same format for years. I have always wanted to publish it in colour, and the advent of Lightning Source's budget colour range has given me the chance to do just that.

I also wanted to publish a digital edition. In fact, I wanted to publish digital editions of *all* our books, but - as so many of our books and magazines are image led - this was nowhere near as simple as it sounds. It took my dear nephew, colleague and friend Dave Braund-Phillips something in the region of eighteen months to work out how to do it satisfactorily for all the different platforms. We also had to decide which platforms we would publish to (all of them in the end) and who to use to distribute the digital editions of the magazine. We had tried one distributor before, but despite all the tales that we had heard about the many millions of all digital products that have been sold, we had sold three in about two years, and it was costing us money every month. So we had to do a lot of research before deciding to accept the kind offer of help from our friends at 11th Dimension Publishing.

And finally, I also wanted to have a major rethink about how the magazine worked, and what was in it. There was nothing at all *wrong* with the way that we used to work, indeed I am very proud of the first 49 issues. But various people have moved away, and others have come on board, and - it has to be said - I have learned an awful lot more about magazine design in the past few years. *Animals & Men* has been published intermittently since 1994, but it has basically been a fanzine. In the past few years we have become a professional book publisher whose output is - I believe - on a par with many long established and mainstream publishing houses, and I wanted to see if we could give the magazine a facelift, and a rethink and drag it into the Digital Age, as a proper, professional and readable magazine. I hope that you will agree with me when I say that we have - I believe - succeeded.

The Great Days of Zoology are not done!

So that explains where we have been. Now let's change the subject a bit, and see where we are going.

When I started *Animals & Men* nineteen years ago, it was the only conduit for cryptozoological news to the English-speaking public. Now that is no longer the case. The Internet has changed everything, and there are hundreds of websites full of cryptozoological news, several of them owned, administered or frequented by the CFZ and its allies.

This proliferation of available knowledge means that one is forced to re-examine the role of *Animals & Men* in this brave new digital age. What we have decided to do is re-invent ourselves. If you want a list of as many new and rediscovered species as we can find (and believe me, our list is far from exhaustive) check the daily CFZ Newsblog which is administered by Corinna.

If you want a list of Bigfoot sightings, try the CFZ-owned but autonomous Bigfoot Forums. You will receive a warm welcome there, and other CFZ-related websites present other aspects of the cryptozoological news in - we hope - a sober and unbiased manner. What we are doing here in the new-look *Animals & Men* is to give free-rein to our inner journalists; to look in more depth at the stories that we believe matter, and to try and sort out the wheat from the chaff.

I am very proud to have been pivotal, together with my friend and colleague Dr Karl Shuker, in launching an academic, peer reviewed journal last year. *The Journal of Cryptozoology* will, I sincerely hope, establish a benchmark of scientific rigour for the cryptozoological community, but - I feel - that there is still room for a readable, authoratitive, but above all *enjoyable* magazine for those people who, like I have for nearly half a century now, found the quest for ethnoknown unknown animal species the most fascinating thing in the world.

However, that is not *all* that this magazine is. The organisation that I founded twenty-one years ago is called the Centre for *Fortean* Zoology, and is not - and has never pretended to be - purely a cryptozoological organisation. Cryptozoology is a very important, possibly the *most* important, part of what we do, but it is not all that we do.

If I had my time over again, I might well not have named the CFZ as I did. I have spent so much of the last twenty-one years in trying to explain to people that we are *not* a 'paranormal research group', that we are *not* occultists, and an acquaintance of my wife made her laugh late last year by saying that he had wanted to come along to one of our annual Weird Weekends, but was afraid that we were a cult of some sort.

Sadly not, although I find the idea of being some sort of idiot bastard son of Charlie Manson and the Reverend Moon rather amusing. The CFZ is a very simple organisation with a whole slew of complex beliefs and aims. These include Charles Fort's assertions that there is a lot of damned (as in excluded) data which can not be satisfactorily explained by current scientific thinking. This is not to say that it never *will* be. I, for one, do not believe in hocus pocus and mumbo jumbo, just in the existence of laws of science that we have not discovered yet.

Our other basic tenet of belief is that the world, and western society in particular was a much better place when Natural History was a discipline generally practised by people of all ages, and all walks of life throughout society. I, in particular, feel that there is something rather sinister about the way that successive British governments have made the practise of Natural History as a hobby ever more difficult and marginalised in today's society. We want to change that, and are immensely proud that so many kids have found what we do a door through which they can discover the complex beauty of the world around us. If we continue to do that, and never discover a new species, we shall have achieved something of great value.

And yes, by the way….they [1] *do* owe us [2] a living.

Slainte

1. The people who rule, own and exploit this planet for money and power, and who have no moral imperative, and no feeling of stewardship for those creatures (human or otherwise) who live there.
2. It is not just data that can be damned as in excluded. People and animals increasingly are as well.

Newsfile

Is the Japanese Otter extinct?

The Japanese otter (*Lutra nippon*) was - according to official records - last seen in the wild in 1979, along the Shinjogawa river in Susaki, Kochi Prefecture (left). During the summer of 2012, the Japanese government declared the Japanese otter (which according to the IUCN is a subspecies of the Eurasian otter [1,2]) to be extinct. In an irresistible parallel to the situation that occurred when the Obama administration declared the Eastern puma to be extinct (see *A&M*49) there have been a whole slew of sightings in the immediate wake of the decision. There have been fifteen to date, all in Ehime Prefecture (right)

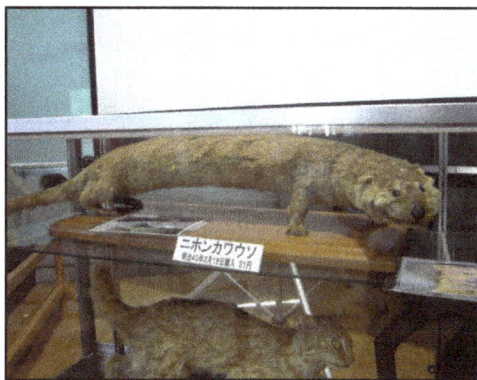

places beyond our knowledge. That was the case when we rediscovered the coelacanth and the kunimasu[3]"

Professor Motokazu Ando, 62, of the Tokyo University of Agriculture, who was involved in drafting the documents to designate the Japanese otter as extinct, dismisses such suggestions. *"Wanting it to be alive is not the same thing as scientific fact,"* he said. *"It is inconceivable that active otters with a length of 70 centimetres and a weight of 5 or more kilograms could live in seclusion and go unseen for more than 30 years."*

Taisuke Miyamoto, 50, director of the Nature Kikaku

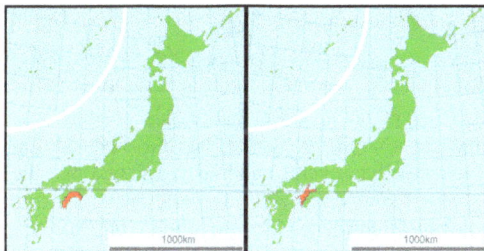

The national Environment Ministry and the Kochi prefectural government carried out a series of surveys across the prefecture from the 1980s to the 2000s but without finding any traces of the otter.

"We made the designation according to the criteria and based on the opinion of experts," an Environment Ministry official said.

However, Ehime Prefecture's natural conservation division disagrees with the assessment. *"There haven't been enough surveys in our prefecture. We cannot determine that it is extinct,"* an official said. Yasunori Miyauchi, 62, the former chief technician at Tobe Zoological Park of Ehime Prefecture, is reported as having said: *"Wild animals can live in*

1. Ruiz-Olmo, J., Loy, A., Cianfrani, C., Yoxon, P., Yoxon, G., de Silva, P.K., Roos, A., Bisther, M., Hajkova, P. & Zemanova, B. (2008). *"Lutra lutra"*. *IUCN Red List of Threatened Species. Version 2009.1*. International Union for Conservation of Nature.
2. The IUCN website states: "Imaizumi and Yoshiyuki (1989) considered Japanese otters a distinct species (*L. nippon*). It is treated separately by Wozencraft in Wilson and Reeder (2005), but that approach is not followed here pending further review".
3. *Oncorhynchus kawamurae*, the black kokanee, or kunimasu in Japanese,is a Japanese taxon of salmon which was thought had gone extinct in 1940, but was discovered to still have a living population in 2010. The fish was thought to have become extinct in 1940 when a hydroelectric project made Lake Tazawa, its native lake, more acidic. Prior to that, 100,000 eggs were transferred to Lake Saiko, which is located about 310 miles (500 kilometers) south of Lake Tazawa, but this attempt to save the fish was thought to have been unsuccessful. However, in 2010, a team of researchers found specimens there.

Wildlife Research Office, says: "*After news of their extinction broke, people who had forgotten about the otter were reminded of it, and the news paradoxically led to an increase in the number of reported sightings. I think some people love the otter, which is the official prefectural animal, while others may be dreaming about or romanticizing the creature.*"

With respect to Professor Ando, his statement doesn't wash. The Eurasian otter (*Lutra lutra*) was hunted to the edge of extinction in the UK only a few short decades ago, but has made a remarkable recovery. Otters are very cryptic mammals, and it would not seem impossible for such a species to have eluded discovery for a few decades.

As far as Yasunori Miyauchi's testimony is concerned, the problem is that I have no idea whatsoever of the status of a 'chief technician' at a Japanese zoo, nor indeed do I have any idea how important an institution Tobe Zoological Park is. He could be anywhere from a retired maintenance man to a senior zoologist. However, his assertion about the Japanese salmonid is quite correct, even though his claim about the coelacanth is one that has been so ludicrously over-used as to be essentially meaningless. Not for nothing did my friend and colleague Darren Naish once describe the prehistoric fish as a 'red herring'.

The most interesting quote is from Taisuke Miyamoto, who points out that the otter is culturally significant for the people of the area, and goes on to say that in his opinion he thinks that the sightings are of other mustelid species. Unlike in the United States when there was a clear, and very shabby rationale behind declaring the eastern puma extinct, I have not been able to find a political motive in the case of the Japanese otter.

However, the annals of Fortean Zoology are full of situations of the power of wishful thinking. One example that comes readily to mind is that of the Brentford Griffin [4], when a palpable hoax was given life by the wishful thinking of people in the area who so very badly wanted their heraldic animal to be real.

Sad but true. But is this the case with the Japanese otter? We are just going to have to wait and see.

4. McEwan, G., *Mystery Animals of Britain and Ireland* (Robert Hale, London, 1986)

The Return of the Danish Wolf

We are indebted to our friend and colleague Lars Thomas for this next story, which - sadly - doesn't have a happy ending, although it is very indicative of the way that the population dynamics of the larger European carnivores are presently in a fascinating state of flux.

The first that we heard of this story was on the 16th October when Lars wrote [1]: "*Strange things are afoot in northern Denmark. On October 14th, a group of birdwatchers in Thy National Park in northern Denmark saw and photographed what they believed was a wolf*".

Lars went on to write that if this creature was indeed a wolf, it would be the first one in Denmark for several hundred years, and that the "*closest confirmed wolf sighting was a young animal in northern Germany about 100 km south of the Danish-German border a couple of years ago. And that's some 500 km south of Thy National Park*".

There have been suggestions that wolves had recolonised Denmark before. According to Lars there had been: "*..sightings of a wolf-like animal in the area for the last couple of years, 4 sightings that I know of at least, but this is the first time anyone has taken a photograph. There has also been sightings in the area around Silkeborg, a densely wooded area filled with lakes about 200 km further south. These sightings have been going on for many years. I have documented a lot of*

1. On his Cryptodane blog - http://cryptodane.blogspot.com/ - which is part of the CFZ blog network

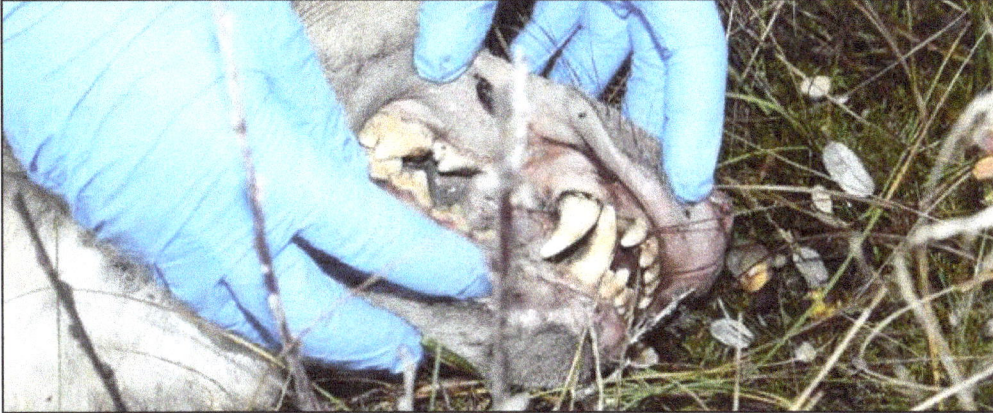

these, including a couple of cases where people have heard wolves (yes - plural) howling".

The next chapter of the story was fairly predictable; Danes reacted to the possibility of a new carnivorous animal in their midst in much the same way as we have seen over the years with the British (on the subject of alien big cats, or the reintroduction of the white tailed sea eagle) or the Americans (the spread of pumas into areas where they have not lived for many decades). On the 19th November Lars wrote:

"The possible wolf that showed up in Denmark recently - and in my last post - elicited the kind of reactions you would expect. A lot of people were over the moon about it, talking about how fantastic it would be to have this large predator back in Denmark. And of course the panic mongers started coming out of the woodwork, screaming about what all those treehugging greenie f****** hippies would say when the wolf attacked their child - which it would certainly do, because we know from scientific sources (Little Red Riding Hood and so forth) that wolves are bloodthirsty killers... blah, blah, blah. And could we please send a lot of hunters and policemen out there and kill it as quickly as possible.

Fortunately not everybody lost their head in such an obvious manner. The environmental authorities mounted a whole string of automatic cameras all over the area and supplied the animal with plenty of bait - dead deer and various other juicy pieces of meat - and then sat back and waited. It almost ended in a major stampede, as some started to offer rewards for a definite picture of the animal. Luckily the reward offer was withdrawn before people started galloping all over the restricted area where the "wolf" was seen

Alas poor wolf - the story did not have a happy ending. As can be seen from this link: http://jyllands-posten.dk/indland/article4913999.ece

The animal was found dead a couple of days ago. Everybody feared it had been shot by some vigilante type or poisoned, but it seems the poor animal died from starvation in spite of the large number of camera baits. It is now going to be dissected to determine the exact course of death, and a DNA-test will be performed to see whether it was in fact a wolf, and if so from what population. Stay tuned for further developments".

It wasn't until 7th December (again on his blog) that Lars tied up the loose ends of the story. *"First of all - the animal was not shot or poisoned or anything sinister. The poor thing was suffering from some form of tumour in the chest cavity - not cancerous, but caused by some form of inflammation. The tumour ended up being so large the animal couldn't eat, and finally wasn't able to breathe".*

But it definitely *was* a wolf. Lars finished up: *"It was indeed a bona fide wolf. According to the genetic analysis it must have come from the German population, so it had been on quite a walkabout before it ended up in northern Denmark. Nobody knows for how long it has been here, but probably for at least a year. I have a sighting from the border area with Germany from early 2011 from a girl of 16 who thought she saw a wolf running across the road early one morning when she was on her bicycle driving to school. This could have been the first entry of the wolf into Denmark.*

Anyway - what is certain is that the now dead wolf represents the first confirmed wolf-sighting in Denmark in 199 years! Not bad. I wonder what comes next?"

Us too, mate. My bet is probably lynx. They are in Germany, Poland and Sweden. How long before one crosses the border?

Man Beasts (BHM)

Almasty 'huts' in Crimea

The following pictures were sent to us in mid-January by our old friend Gregoriy Panchenko, the eminent Russian cryptozoologist who lives in the Ukraine. They were taken in the Crimea, in a place where there have been many almasty sightings, and Gregoriy believes that these are

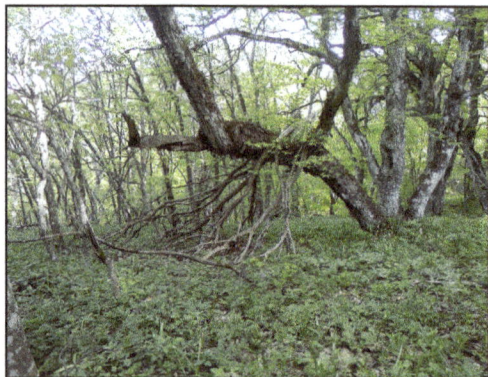

'almasty huts' performing a similar role for the Eurasian manbeasts as so-called 'bigfoot teepees' do in North America.

The Bigfoot Phenomena website [1] describes this phenomenon, and goes on to list a number of different types that have been noted.

Other researchers are less impressed with the phenomenon, believing that whether artificial or man-made they have nothing to do with the man-beast phenomenon.

We are going to sit on the fence for the moment and await developments, although it is interesting to note that we believe that these are the first photographs of structures like this that have been obtained from Russia.

1. http://www.bigfootphenomena.webs.com/types.html

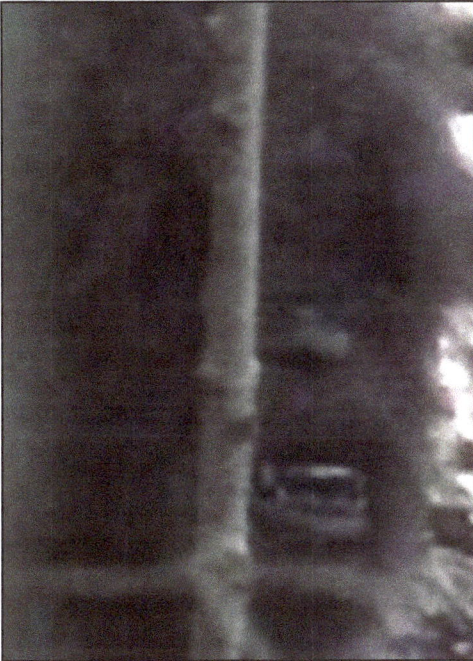

Waltzing Matilda

On page 37 of this current issue you can find a partial list of some recent claims that have emerged from parts of the Bigfoot Research Community. Several of these have appeared at around the same time as the Melba Ketchum paper, and claims and counter claims have been made linking them or distancing them. One claim that certainly *is* linked to the Ketchum paper is the release of a video, attributed to the Erikson Project, which claims to show a sleeping sasquatch.

Some people accept the claims, others reject it out of hand, and yet others claim that it is a film of a sleeping dog. More recently, however, as we were going to press a set of photographs was leaked, and acquired by a researcher called Bill Munn. Munn claims (plausibly enough) that the pictures show nothing more than someone wearing a doctored Chewbacca mask (Chewbacca being a character from *Star Wars*). I don't know who leaked the pictures or why, and I am not going to comment on claims of their provenance, but it has been suggested that they are photographs of an animal nicknamed

'Matilda' by the Erikson Project, who is the same creature seen in the video purporting to be of a sleeping sasquatch. There are also claims that the Erikson Project *rejected* this series of pictures, and other claims that they were on sale by person or persons unknown (in fact I *do* know who it has been claimed had them for sale, but on legal advice I am not going to say). I am finding this whole saga absolutely fascinating, and being very new to any part of the Bigfoot Research Community, I do not know yet who can be trusted, and whose testimony has to be taken decidedly *cum grano salis*. So I am just going to sit firmly on the fence for a while, and I suggest that you all join me and see what happens.

Is this "Matilda"?

The source of the "creature" images never referred to it as "Matilda" but the descriptions I have read about "Matilda" seem to fit what is seen here. To me, it is obvious this "creature" is simply a Chewbacca mask with the hair reworked to be a different color and texture. I welcome comments from anyone who thinks otherwise.

Mystery Cats

The effects of range extension

Various North American felids are extending their range, and it is interesting to note how the human population is dealing with this.

Over on the CFZ Big Cat blog we chronicle all the news stories that we can find of human/big cat interaction as well as cryptozoological or quasi-cryptozoological sightings. If one compares the recent history of interaction between humans and pumas in North America with the recent history of human/tiger and human/leopard interaction in India, I am afraid that the citizens of the First World come off less sympathetically.

In a majority of the reported cases, when a puma has been reported in the vicinity of human habitation it has been shot, whereas a subjective look at the similar news reports from India would seem to indicate that their *modus operandi* is to trap and relocate problem animals. Granted, the two Indian species are protected by law in a way that *P. concolor* is not, and an in depth analysis of the available data may well show a different story.

I invite anyone who is interested in such matters to have a look at the archives on the CFZ Big Cat blog.

Peoria Puma

The above picture was taken by a trail-camera in Peoria, Illinois during January. It is an area where there have been other recent sightings of mysterious felids. It has been suggested that it is a bobcat, but the animal appears to have along ringed tail. Bobcats are possessed of a small stumpy tail. So what is it?

There have been intriguing accounts of what appear to be bobcats with long tails from various parts of the United States for many years. In a book we published late last year, Andrew Gable includes a photo of one killed in the early 1920s; one of a series of animals captured or killed over the years.

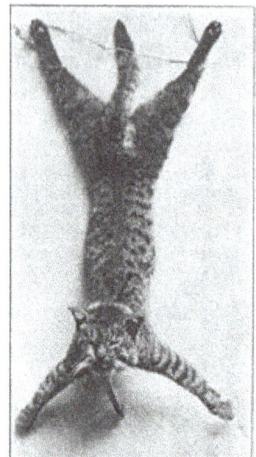

This also appears to have a partially ringed tail. Could it be a similar animal? If so, my earlier question still stands.

A Welsh Rare Bit

In late February I had an email from Gavin Lloyd Wilson, the erstwhile News Editor for CFZ Online. He wrote:

Hi Jon, You may have seen this story already, but just in case...

He included a link [1]. The story began:

For those not aware the Western Telegraph website ran a story this morning about a bizarre creature that had apparently been washed up on Tenby's South Beach on the weekend of the 23rd-24th of February 2013..

The website then included a link to the *WT* story [2].

We started receiving emails about this as the story broke and it is genuinely quite interesting because it is eerily similar to the Montawk Monster/Manhattan Monster. Both creatures are, on initial viewing, very odd looking. Both are virtually hairless, both have strange proportions, both are discoloured, both seem to

have long almost human-like fingers and both seem to have a beak/nose not usually associated with animals of this size. Could this really be evidence of Alien life in West Wales?

The article was from a website run by a paranormal group called Pembrokeshire Beyond who certainly seem to have their heads screwed on, because they go on to say:

Without further examination, which would be impossible as the remains were moved by people who appeared to be council workmen, we can't be 100% sure what it is. But I do not believe it is in any way extra-terrestrial in origin, nor do I believe that it could be the result of any bio-experiment gone wrong.

The Montawk Monster is believed by some to be a decomposing Raccoon, I believe that something similar has happened here. An animal the size of a small Dog (looking at the teeth it could even be a Dog) has fallen into the waters of the Bristol Channel just before or prior to death. In the waters of the Channel it has begun to decompose which has caused the skin to discolour and rot around the snout/nose giving an almost beak like appearance, the fur

1. http://www.pembrokeshirebeyond.co.uk/the-tenby-thing/
2. http://www.westerntelegraph.co.uk/news/county/10249810.Can_you_solve_the_mystery_of_the_Beast_from_the_East_/?action=success

to fall out which is making the animal appear almost unrecognisable upon first glance (although there are hints of white fur on the creatures back) and the body itself has swollen due to internal gases caused by the breakdown of organic matter.

In a world were such things quite often get blown out of all proportion by people who really should know better, it is refreshing to find such a sensible account. I wrote to them:

Hi Guys,

I am passing the picture over to one of our zoologists to identify. My guess is dog. However, I thought I can do a couple of things to help. Firstly, you have mis-spelled Montauk on the graphic. Its an easy thing to do. And secondly, there is photographic proof that the MM was a dead raccoon. Check this out: http:/orteanzoology.blogspot.co.uk/2009/06/ montauk-monster-revealed.html

But well done on correctly describing it as a dead animal made hairless by immersion in seawater rather than jumping on the 'alien' bandwagon. There are too many people these days prepared to make capital out of stories like this. I am glad to see you guys don't.

Jon Downes,
Director, Centre for Fortean Zoology

POSTSCRIPT: Lars Thomas confirms it is a dog but because the picture is not terribly good he can not identify it to breed. Max, who is terribly busy, wrote back that it is probably a dog.

Richard Freeman thought that it looked like the corpse of a badger, but all are agreed that it is a terrestrial carnivoran.

So no alien then?

Over the Sea to Ireland

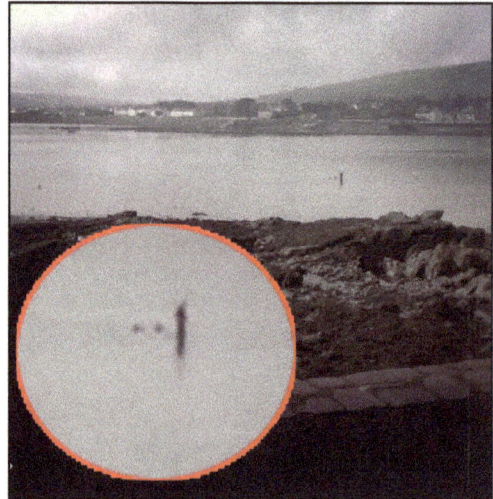

In mid January, I received the following email and the above picture from Treasa Murphy, an acquaintance of mine who works for Irish radio. She wrote:

Dear Jon,

This picture was taken by a local man at Renard Point, Cahersiveen close to Valentia Island at 8.30 am last Sunday. The photographer said he noticed it in the water, took a photo and then it disappeared shortly afterwards. It was around 50 yards from shore.

Thank you for this, hope you're well.

Kind regards,

Treasa.

I agreed to pass it around the great and the good of cryptozoology, but it was Corinna who came out with the most plausible explanation: that it was a fisherman wearing waders and with an elongate floating keepnet thingy. And that the reason the creature had disappeared is that the fisherman was hungry and went home for his breakfast!

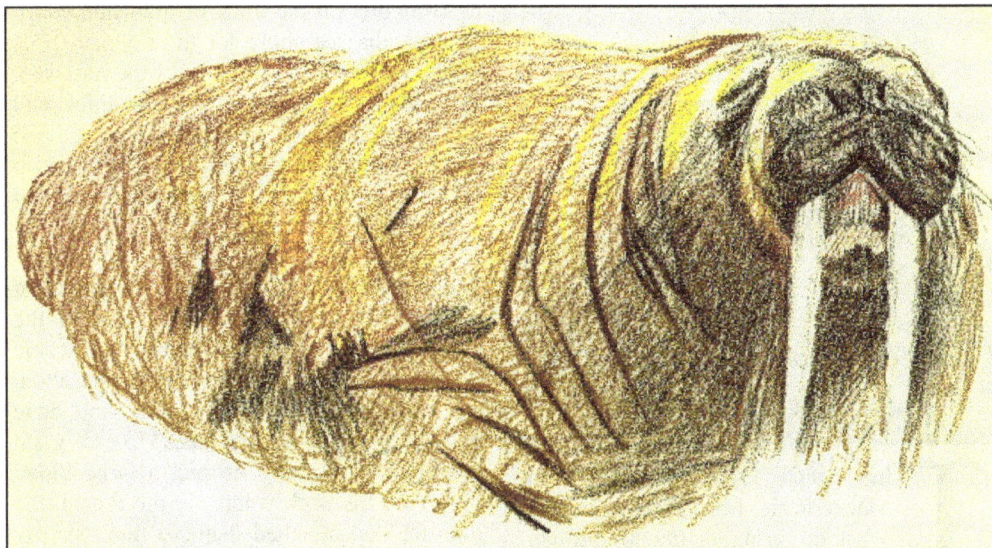

I am the Walrus

Glen Vaudrey reports:

Early March saw a walrus turning up in Orkney on the island of North Ronaldsay, a rare enough visit in itself as the walrus was considerably further south than would be expected. The animal was discovered by Mark Warren, an assistant warden at the North Ronaldsay Bird Observatory who described the discovery:

> "I was walking along the beach looking for birds, basically. At first I thought it was a dead whale so I threw a stone at it and it woke up."

A great way to greet an out of place animal.

The sighting was first reported by the BBC *Scotland News* website on 3/03/2013 but the animal's stay was rather brief as it was observed to be heading back out to sea on the 4/03/2013. It was the first walrus sighting in Orkney since August 1986 when one was spotted on the island of Eynhallow.

However there was a report of a sea serpent washing ashore on Orkney in 1894 that could well have been the remains of walrus, as can be seen in the following tantalising article from the *Gloucester Citizen.*

Gloucester Citizen *18th September 1894*
The Miscellany

In default of the sea serpent, this will do. A Kirkwall correspondent says some sea monster has been washed ashore at Traittland Ronsay.

It resembles a walrus, is from ten to twelve feet long, and in a decomposed state. The back of the neck is like that of an elephant; the snout resembles a pig and on the body is greyish hair. Another correspondent says it is thicker than a horse, but is probably much wasted by decomposition. The colour is a light grey. No one seems to be able to say what the species belongs to.

Sources
- BBC Scotland News website 03/03/2013 & 04/03/2013
- *Daily Record* 04/03/2013
- Vaudrey,Glen *Sea Serpent Carcasses Scotland: from the Stronsa Monster to Loch Ness* (CFZ Press 2012)

The Czech cryptozoologist Ivan Mackerle is best known in the West for bringing the Mongolian deathworm to the attention of the public. It is likely that this fascinating cryptid would have still been as obscure as ever if it were not for Ivan. But this intrepid man did far more than hunt for the deathworm.

Ivan grew up in Prague and was fascinated by exploration from an early age. Despite scouting being banned under communism he formed a scouting club with some friends. Dubbed the Fast Arrows they had a secret clubhouse equipped with a radio for Morse code. He collected clippings about cryptids and later became interested in other areas of forteana such as forbidden archaeology and ghosts.

He studied engineering, and after university he worked as a designer. He married and had a son, Danny, and after the fall of communism he was free to take up his childhood passion of exploring. His adventures took him to Madagascar in search of the elephant bird and the man-eating tree; to Mongolia on the track of the deathworm; to Australia searching for the thylacine; Sri Lanka searching for remains of the nittaweo; to South America looking for lost cities; and the Siberian Tiga trying to unearth odd underground dwellings. He was also involved in a Czech TV show 'Enigmas and Mysteries'.

In one weird adventure Ivan was warned that the deathworm was a baleful spirit of the Gobi that struck down those who came looking for it. Ivan had a vivid dream about the monster exploding out of a sand dune and clamping onto his back like a giant leech. Next morning he had a large blood blister on his back where, in the dream, the monster had attached itself to him. Shortly after he had a heart attack. Heart problems were to beleaguer him for the rest of his life.

Ivan's work in Mongolia was inspirational to both myself and Adam Davies in our own quests for the deathworm. I met him Ivan at the *Fortean Times* Unconvention and found him to be a fascinating and charming man. He kindly gave me some DVDs of his expedition films.

Ivan passed away on January 3rd 2013. This man, the stuff of *Boy's Own* adventure and old school exploration will be sorely missed. We need more of his kind in a world increasingly filled with armchair experts and sceptics who never dare the jungles, mountains, swamps and deserts.

Ivan's torch has been passed onto to Danny who accompanied his farther on many expeditions. Danny is now planning to search for a mystery primate in Madagascar, and the ropen of New Guinea. **RF**

OBITUARY

Carl Marshall's Column

L ong before I discovered cryptozoology I had always had an interest in our planet's more extraordinary wildlife, whether it's the mysterious strepsirrhine primates like the aye-ayes of Madagascar, with their long slender fingers, big eyes, and large sensitive ears, looking rather like some demonic Mogwai, or the incredibly noxious bombardier beetles, ejecting a near-boiling chemical spray from their abdomens at potential attackers. Incredible creatures like these have always fascinated me, and the available literature about such strange wildlife greatly influenced my interests during childhood, moulding the person who I am today. Incredulous as these creatures may seem, natural obscurities such as these are not always restricted to the damp impenetrable rainforests of the world or the elevated mountain tops enveloped in thick rolling clouds; to find zoological anomalies to quench our exploratory thirst sometimes something atypical can be found much closer to home, exemplified by the discovery illustrated below.

These curiously pigmented common moles *Talpa europaea* were both killed in mole traps at the same location in an agricultural field on the outskirts of a wood in Newbury, Berkshire and sent [by the farmer who wishes to remain anonymous] to a friend of my father, a fellow taxidermist based in Wales, Mr. Martin Bennet. Martin mounted these exceptional Talpids, and has assured me that they are not taxidermic forgeries; they arrived at his workshop in this exact condition and he has not altered their appearance in any way. On first examination I

Newbury's Mutant Moles

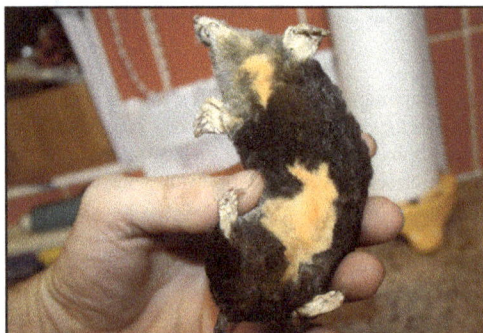

thought they could be partial (mosaic) albinos. However they were very vividly coloured, and my initial investigation to identify this mutation was inconclusive, so I decided to contact Dr. Karl Shuker, an expert in teratological enigmas. After talking with Dr. Shuker, it seems more likely that they are actually exhibiting flavistic mosaicism; two rare phenomena where an unusual golden orange mutation (flavism) has developed only in patches (mosaicism ie. patches of abnormal golden pigmentation amid an otherwise typical pelage).

Dr. Shuker also informed me that he had not seen comparable specimens like these before, and confirmed my theory that they were probably closely related, sharing the same genetic makeup for this mutation, and were possibly siblings. Dr. Shuker also informed me that the European mole actually yields quite a wide range of colour mutations, but most people are unaware of them due to the subterranean nature of this family.

See below for a brief list of recognised colour morphs:

- Albinoism - lacking melanin. Also known as amelanistic. White body (ie. lacking melanin) with pink/red eyes (ie. lacking melanin in the cornea).
- Dilutinic/Maltesing - washed out colour eg. black becomes blue (grey).
- Erythristic - black pigment is converted to red.
- Flavistic - golden mutation.
- Hypermelanistic - having excessive black and/or brown pigment.
- Hypomelanistic - having less black and/or brown pigment.
- Leucistic - similar to albinos, white, but having dark eyes.
- Mosaicism - abnormal patches.
- Piebold (mutation) - white patches.

It's also safe to presume that there will be more of these moles displaying this cryptic colouration at this hidden location in Newbury. If any interested researcher wishes to investigate these moles further please contact carlmarshall83@live.com or the Centre for Fortean Zoology.

I would like to thank Mr. Martin Bennet for letting me have these magnificent mutant moles, and also I would like to thank Dr. Karl P.N Shuker for his professional help. I would also like to take this opportunity to thank my good friend Mr. Jonathan Downes (Director of the Centre for Fortean Zoology) for all his help and encouragement, and for welcoming me into the CFZ family.

Carl Marshall works at Stratford Butterfly Farm and is a fine field naturalist. Over the past couple of years he has become a very enthusiastic member of the CFZ, and his quasi-fortean view of British natural history fits in perfectly with my own. He was, therefore, the perfect choice as a columnist for the brave new *Animals & Men*, and we are proud to have him aboard.

One of the most frequent questions I am asked is 'what is the largest spider in the world?' the answer as many people do know is the Goliath Bird Eater, *Theraphosa blondi* (opposite) found in South America in countries such as Venezuela, Brazil and French Guyana. They are generally referred to as 'dinner plate' size. It's the heaviest spider for sure but actually it does not have the largest leg span – that belongs to one of the huntsmen spiders *Heteropoda maxima* (see below).

This arachnid was apparently only described in 2001 from a specimen in the *Muséum national d'Histoire naturelle,* Paris. It resides deep in caves in Laos and I guess discovering it will have given someone quite a fright. But get this, the leg span can be anywhere from 250-300mm. Get your ruler out and measure it, that's a foot long spider!!

I am seriously impressed and of course I would love to go out there and find a specimen, but that's on my bucket list.

I began fantasising. Somewhere out there is a giant tarantula of 'Shelob' proportions but where would I have the best opportunity to find it?

Welcome to the mythical beast that is J'ba Fofi (giant spider) said to be stalking the forest of the Congo. It is considered to be a cryptid and various people testify to have seen it but sadly I am not one. It is supposed to have the same brown body as that of a tarantula with a leg span of some six feet. Who feels insignificant now Mr Huntsman? The juveniles are meant to be bright yellow with purple abdomens, turning brown with each moult.

A little rather lazy research on the decidedly dodgy Wikipedia found accounts of sightings of a giant spider – but not just from the Congo.

1938: Congo
Reginald and Margurite Lloyd were driving a Ford truck through a trail when they spotted a spider resembling a tarantula crossing the path ahead of them. The creature's leg span was estimated at 3 feet. Their daughter Miss Margurite Lloyd would later relate this story in the 1990s to William Gibbons.

1942: Papua, New Guinea
An Australian soldier at the Kokoda Trail said that he

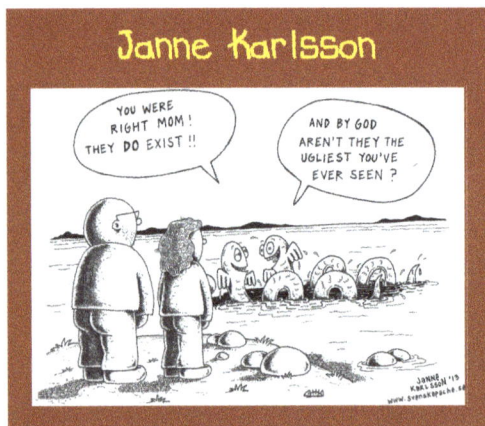

Janne Karlsson

encountered a puppy-sized spider inhabiting a 10 to 15 foot sized web. It was described with having a bulky body, was black, and hairy like a tarantula.

1948: Leesville, Louisiana, United States
William Slaydon and his grandchildren were walking north on Highway 171 to church when he motioned them to stop. After hearing a rustling in the bushes ahead a spider described as being the size of a washtub emerged and crossed the road. One of the grandchildren would later tell this story to his own son, Todd Partain, director of the documentary film "Eyes In The Dark: The Sasquatch Experience."

2001: Cameroon
Timbo, chief of the Baka tribe in Cameroon tells William Gibbons that in November 2000 a J'ba Fofi had built a nest near their village.

2011: Amazon
British cinematographer Richard Terry travelled to the Amazon to investigate reports of giant spiders in the June 13th episode of Man v Monster. At a remote village he was informed that giant spiders lived in holes deep within the jungle and measured roughly four feet in diameter.

2013: Texas
A report from Texas comes of an "abnormally large spider" which was found crawling on the side of a house. It measured about four feet in diameter. It reportedly took several men several shots to kill the creature.

Should we dismiss these anecdotal sightings as nonsense, after all there's no photographic evidence to support them? I am no biologist but I am sure that no spider of that size would be able to survive on planet earth for many reasons. Yet maybe there are some huge arachnids out there from a bygone age roaming around, hiding from man and surviving. There is fossil evidence of one metre long scorpions and more with sea scorpions so why not huge spiders?

There's one way to find out of course and that's to get my exploring trousers on and visit the Congo. Yes, that's another one for the bucket list. I don't fancy getting kidnapped just at the moment. Well J'ba Fofi if you are out there good luck to you. I dream of finding you, I think about how wonderful; how magnificent you might be. As David Bowie once sang, 'one day though it might as well be some day, you and I will rise up all the way, because of what you are – the prettiest star.

Born in Birmingham, England Carl Portman has always followed the maxim 'interest is where you find it' and this certainly applies to natural history. He has bred endangered species of tarantula spiders, written two books on natural history travel and lectures around middle England on animals and rainforests. Oddly he has a diploma in sexing juvenile theraphosid spiders, is an English Chess Federation County Chess Master, supports Aston Villa and has a strange addiction to Turkish Delight (covered in chocolate). Having worked for the Ministry of Defence for 30 years he now spends his time doing lecturing, chess coaching, some photography and management consulting.

He has spent time studying animals in the rainforests of Australia, Ecuador and Costa Rica searching for new and ever curious insects and arachnids and has a desire to find a new species somewhere in the world. His motto is 'Don't complain about the dark, light a few candles'.

He is married to Susan and lives in Oxfordshire. Their two Border Collies, Darwin and Dickens keep them fit and ensure that there is never a dull moment in the household.

W ell, as 2012 ended and a new year dawned (yep, in case you didn't notice, the Maya got it spectacularly wrong!), quite a lot was going on with regard to the CFZ here in the United States. At the end of the year, CFZ Press published my book, *Wildman! - The Monstrous and Mysterious Saga of the British Bigfoot*.

This is a book I have wanted to write for years, and finally got around to it. Basically, as its name suggests, *Wildman!* is a full-length study of Bigfoot, feral people, and wild man-type entities in the UK dating back centuries. And at around 120,000 words, it's the longest book I've written since my first book, *A Covert Agenda*, was published back in the UFO-saturated year of 1997.

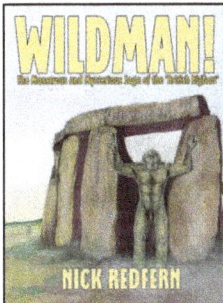

And here – courtesy of the good Mr. Downes – is the blurb for *Wildman!*

> *'The huge forests of the United States are home to Sasquatch. The Abominable Snowman roams the Himalayas. Australia has a similar beast, the Yowie. In China there lurks a giant, bipedal creature called the Yeren. From the Caucasus Mountains in Eurasia stories of the Almas circulate. And then there's the highly controversial matter of Bigfoot in Britain. For years, Nick Redfern has been on the trail of this mystifying monster of the British kind - one that provokes fear, amazement and controversy whenever it rears its horrific, hairy head. The Shug-Monkey, the Beast of Bolam, the Big Grey Man, the Man-Monkey, and the Wild Man of Orford are just a few of its many names.'*

Linda Godfrey, the author of many werewolf-themed books, including *Real Wolfmen*, *The Michigan Dogman*, *Werewolves*, *The Beast of Bray Road*, and *Hunting the American Werewolf*, has done an excellent review of *Wildman!*, which reads as follows:

> *'Bigfoot in the British Isles? The idea of huge, hairy man-apes hiding out in the manor-dotted, bucolic countryside of the British Isles seems as likely at first blush as the queen switching out her Earl Grey for chugs of Red Bull. But Nick Redfern's new book, Wild Man! The Monstrous and Mysterious Saga of the British Bigfoot, (CFZ Press, 2012) makes an exhaustive and surprisingly compelling case that people have indeed been seeing anomalous, Bigfoot-like creatures from Shugborough to Derbyshire, perhaps for centuries!*

> *'Redfern has to be one of the most prolific current writers on strange and cryptozoological topics -- he releases books faster than I can find time to review them. But I wanted to make sure that I got this one posted because it covers so many topics of interest to the worldwide cadre of Bigfoot-seekers. It has the additional advantage of being authored by a native of the British soil who possesses many area connections and much firsthand knowledge.*

> *'Redfern starts by grounding the wild man topic in medieval history, harking back to the wodewose, or hairy, naked men seen throughout the forests of England since at least the early 14th century. But he notes that the wodewose was described as very close to human underneath all that fur.*

> *'The nearly 300, large pages are packed with anecdotes in a timeframe that ranges from ancient to contemporary, but the book is about more than mysterious case studies. Redfern doesn't shy away from possible explanations for the anomalous creatures,*

venturing into possibilities such as misidentifications with primates released from the private zoos of eccentric collectors.

'He also notes the difficulty of explaining how large predators could subsist in the English countryside without detection and lays out a theory that they may be "semi-physical" or products of some other process that is not presently known to science.

'Whatever these wild things are, Redfern's examination of them belongs on the bookshelf of anyone interested in anomalous creatures, especially those of the hairy hominoid kind. There may be more beasties hiding in that famed British mist than anyone has previously guessed.'

And there ends Linda's review.

Coincidentally, at the same time that *Wildman!* was published, Patrick Huyghe's Anomalist Books published another cryptozoological title from me: *Monster Diary: On the Road in Search of Strange and Sinister Creatures*. And here's the blurb for that book from Patrick:

"Monster Diary *reveals that many of the unknown animals of our planet are not all they seem to be. They may appear to be flesh-and-blood creatures, but is that what they really are? Redfern sets out to prove that the true nature of the fearsome creatures that dwell in dark and shadowy woods, atop imposing mountainous peaks, and within the depths of murky lakes and rivers can only be understood with a knowledge of ancient rituals designed to conjure up foul life forms from some terrible realm, ominous sacrificial ceremonies undertaken in the dead of night, and disturbing occult rites. Monsters do exist. Monsters are among us. But they are not what you probably think there are."*

Moving on, Saturday, January 26, 2013, saw me heading down to Austin, Texas – with Cryptomundo's Craig Woolheater - to speak at a one-day gig at the *Museum of the Weird*. If you ever

get the chance, check out the museum. If monsters and horror movies are your thing, you'll love it.

There's a life size model of the beast made famous in *Creature from the Black Lagoon*, a two-headed calf, a mermaid (of course it's real!), and much more of a creepy-creature nature. And there's an excellent shop where you can buy all manner of monster- and horror-driven t-shirts, posters, postcards, badges, and much, much more.

There's also a life-size 'gripping hand' of King Kong that you can stand inside of and have your picture taken, while an equally life-sized head of the beast looms over you!

I was lecturing on the subject of my *Wildman!* book,

namely Bigfoot in Britain, with most of the presentation focusing on the infamous Man-Monkey of the Shropshire Union Canal. Also speaking were good friends Ken Gerhard (*Big Bird!*, *Monsters of Texas*, and the forthcoming and much anticipated *Encounters with Flying Humanoids: Mothman, Manbirds, Gargoyles, and Other Winged Beasts*) and Lyle Blackburn, author of *The Beast of Boggy Creek*, which is – beyond any shadow of doubt - the definitive study of the hairy monstrosity made famous in the 1972 film, *The Legend of Boggy Creek*. Also while at the museum, I met up with (for the first time) David Coleman, author of *The Bigfoot Filmography* book – which is most definitely essential reading for anyone wanting a comprehensive study of Bigfoot in films and TV. David is a cool chap and I recommend his 'Filmography' – about which he spoke at the gig - to one and all. The event was well attended, and was followed by dinner and drinks. In other words, a great time was had by everyone.

On the weekend of February 16/17, I found myself (as you do) trekking around the woods near Ennis, Texas with good friend, and Bigfoot-seeker, Larry Parks. For years, there have been reports of large, hairy, man-like creatures in the area, so we figured let's go and have a scout around.

Ultimately, Larry and I failed to find any evidence of Bigfoot, as we fought our way through the dense woods that overlook the wild waters of the Trinity River. We did, however, find lots of cool stuff in the woods: deer antlers and bones, a large turtle shell, an impressively-sized spider egg sac, and an abandoned old school bus that was home to a huge, black, and decidedly sinister-looking, turkey vulture.

And there's more to come this year from me: an expedition in search of the legendary 'Oklahoma Octopus', another book (*Monster Files*), and several lectures and conferences where I'll be flying the flag of, and promoting, the good ship CFZ.

Nick Redfern can be contacted at his *World of Whatever* blog:
http://nickredfernfortean.blogspot.com

TROLLS: THE MAN BEASTS OF NORTHERN EUROPE

Trolls loom large in the legends of Scandinavia. Hulking, hairy, man-like beasts, shaggy dwarfs or human-like beasts with cow tails; there are many stories about them, some from surprisingly recent times.

The most popular depiction of a troll is a hairy, wild, man-like creature of huge strength but lacking in wits. Females have long, drooping breasts. There are stories of troll / human hybridisation.

Lars Thomas, whilst researching a forthcoming book on the subject, was struck by the similarity between the ancient descriptions of trolls and the more recent descriptions of the almasty, the Russian man-beast. One account of an old Danish king who loved to hunt says that his favourite quarry were trolls because of their hardness to catch and savagery when cornered. The description given sounds very like the eyewitness accounts I have heard from the Caucasus Mountains of almasty. RF

watcher of the skies

CORINNA DOWNES

The far south coast off New South Wales yields rare birds

A pair of fluffy Gould's petrel – a very rare seabird - was found by the National Parks and Wildlife Service on Montague Island - near Narooma, off the New South Wales south coast in early February. Dr Amy Harris, Shorebird Recovery Officer, said: *"It's a great discovery! We have one pair on the island which have a chick at the moment,"* she said.

Source: http://www.abc.net.au/news/2013-02-27/rare-birds-found-off-far-south-coast/4542934

Critically endangered New Zealand storm-petrels - crucial discovery of breeding ground

For 200 years up until 2003, the sparrow-sized New Zealand storm petrels were thought to be extinct. However in that year several were spotted out to sea and now – for the first time – their breeding location has been discovered, just 50 km from Auckland City, situated on Little Barrier Island, Hauturu in the Hauraki Gulf Marine Park. Chris Gaskin, who is the Important Bird Area Programme Manager for Forest and Bird (BirdLife in New Zealand) and Dr Matt Rayner from the University of Auckland are the leaders of a team of researchers that has been using radio receivers to zero in on the breeding site of this critically endangered seabird.

Specially designed net guns and small 1g radio transmitters were fitted to the 24 birds caught at sea this year. When the bird was stationary at night, team members were then able to get a clear fix on their whereabouts. *"The site being monitored is very fragile and with birds at a delicate stage in their breeding cycle. We are using automated equipment for the most part and maintaining a hands-off approach, although team members visiting the vicinity have also been keeping watch,"* explained Dr Rayner.

Source: http://www.wildlifeextra.com/go/news/nz-petrels.html

Indonesian new owl species is formally described

Back in September 2003 a new owl species was discovered in Lombok, Indonesia but it was not until recently that it was formally described, according to a BBC News item dated 14th February this year.

Known as the Rinjani scops owl (*Otus jolandae*), this "common" owl is the first endemic bird species recorded on the island of Lombok, and the first study of this owl, by an international team of scientists, is published in the journal *PLoS One*.

"*I found the new owl on 3 September 2003, and Ben King found it independently at a different location on 7 September 2003,*" explained George Sangster, lead researcher from Stockholm University's Department of Zoology in Stockholm, Sweden, describing his first encounter with the new species.

Rare bird rediscovered after 83 years

The Sillem's mountain finch is one of the world's rarest birds, and was first spotted in 1929 by Dutch ornithologist Jerome Sillem in Aksai Chin, Xinjiang. It was not until June last year that one was seen again, this time in the Yenigou valley, Qinghai, 1,500 kilometres away.

Writing on his blog, Yann revealed that he '*did not venture into the upper reaches of Yenigou Valley... with any scientific goal, and it was not a*

birdwatching nor a photography expedition either'. He added: "*After setting up camp, there was still time to explore the surroundings and that's when I had the first sighting, and photography, of a finch that I had not seen before. It was sitting quietly among a loose flock of what I identified as being female Tibetan rosefinches* (Kozlowia roborowski)*, a bird that I had not seen before either, but that was on my short list of 'birds to look for' during the trek*".

Source: http://www.scmp.com/news/china/article/1068009/rare-bird-rediscovered-after-83-years

Discovery of Critically Endangered South American Bird in New Area Offers Increased Hope for Avoiding Extinction

The new discovery of a population of the critically endangered Royal cinclodes by the Peruvian environmental group Asociación Ecosistemas Andinos (ECOAN), is providing some increased hope that this bird may be able to be saved from extinction.

During August 2012, ECOAN biologists spotted a Royal cinclodes and possibly a second bird inside the Huaytapallana Regional Conservation Area in Peru's Junín department (comparable to a U.S. state).

The sighting was 29 miles north of the nearest and previously northernmost population discovered in Junín in 2008. "*There may well be fewer than 250 of these birds left in existence,*" said Constantino Aucca Chutas, President of ECOAN. "*These new sightings are therefore quite significant because they raise the odds that this rare species might be saved.*"

Sources:
http://www.surfbirds.com/community-blogs/blog/2012/10/16/discovery-of-critically-endangered-south-american-bird-in-new-area-offers-increased-hope-for-avoiding-extinction/

http://www.abcbirds.org/newsandreports/botw/royal_cinclodes.html

Giant Snakes of the World

Carl Marshall

This article is not to be taken as a definitive coverage of outsized snakes of the world; rather as just a few examples of some giant cryptozoological species, as well as some personal speculation that at least one known species may have possible sub-specific variations of even larger proportions.

During my six years as curatorial assistant at Stratford-Upon-Avon Butterfly Farm I have had the opportunity and privilege to work with many unusual and often deadly species. You see, it's not all pretty butterflies; our latest "nasty" is a Brazilian wandering spider; *Phoneutria sp.* (the murderess) and she really is evil.

We also get donated snakes, usually about eight - ten a year; Colubrids, Boids and pythons – some of which are brought in by the RSPCA due to neglect by owners who are ignorant of the care required, or lack the sensitivity and commitment. The owners themselves who have had to give up their pets, often because the snake grew bigger than they were expecting, also sometimes hand them in.

The largest snake I have personally worked with was a 15 ft (4.5 metres) reticulated python, which - despite its large, but certainly not giant size - was very difficult to deal with because of its immense strength and aggressive nature. So now we pose the question: what happens when an owner buys a potential giant like a reticulated python as a "first snake" and it rapidly grows into a hyper-aggressive monster? Pet shops do sell gigantic snake species like anacondas and reticulated pythons to individuals who (at the time of writing) do not need any specific licensing to buy or keep them. So again, what happens when a snake exceeds 20ft (6 metres) in length and has the appetite to match? The owners have four choices open to them.

They can either:
a) Adapt, and keep the snake
b) Sell or relocate it to a responsible carer
c) Have euthanasia performed by a qualified veterinarian
d) Illegally release it into the wild.

Unfortunately, the latter has happened around the world where there are many instances of invasive snake species damaging the ecological balance of the area that they have escaped to, or been released into. Currently the Florida Everglades in North America has become home to the Burmese python, which is obviously not its country or even continent of origin, and subsequently they are causing all kinds of problems to local wildlife and their habitats.

It is believed by established zoology that the largest species of snake in the world in terms of bulk is the green anaconda from South America, which we know can reach a massive 28-30ft (approx 9.1 metres) long, although some acknowledge 37ft (11.2 metres). But does this species, or any other, grow even bigger? I personally believe the green anaconda *Eunectes murinus* does, possibly up to about 50-55ft (approx 16.7 metres) and also the reticulated python *Python reticulatus* which I feel could, if undisturbed, reach lengths of up to about 45-50ft (approx 15.2 metres) and the following article will discuss the implications of these estimations.

SOUTH AMERICA

Quoted below is a witness account by adventurer F.W. Up de Graff and his team in 1923 regarding a colossal green anaconda in its natural habitat:

""There's a dead alligator over there; let's get out of here."

I turned to look in the direction in which he had pointed. In a moment I saw his mistake. There lay in the mud and water, covered with flies, butterflies and insects of all sorts, the most colossal anaconda which ever my wildest dreams had conjured up. Ten or twelve feet of it lay stretched out on the bank in the mud; the rest of it lay in the clear shallow water, one loop of it under our canoe, its body as thick as a man's waist. I have told the story of its length many times since, but scarcely ever have been believed. It measured fifty feet for certainty, and probably nearer sixty. This I know from the position in which it lay. Our canoe was a twenty-four footer; the snake's head was ten or twelve feet beyond the bow; its tail was a good four feet beyond the stern; the centre of its body was looped up into a huge S, whose length was the length of our dug-out, and whose breadth was a good five feet.

I was in the stern where I couldn't reach the rifles, so I called out to Jack to shoot. He reached out for his weapon, but the noise he made in fumbling for the it alarmed the snake.

With one great swirl of water that nearly wrecked us it vanished. The agility with which it moved was absolutely astounding in view of its great bulk, in striking contrast to the one we skinned. When I thought of how the latter's decapitated body had coiled round my legs and nearly broken them in the last contraction of its dying muscles, I wondered what would have happened

to us had that huge beast in its headlong flight taken a turn round the canoe. How utterly helpless the mightiest of men would be in the coils of such a monster!"

*Up de Graff. F. W.
Head-hunters of the Amazon, 1923.*

A snake of 60ft (18.2 metres)? I personally think a snake of this size is possible, yet improbable. There are even reports of anacondas of 150-200ft (approx 60.9 metres) but these are reports from frightened natives and non-scientific observers, and should not be accepted at face value without sufficient corroborating evidence. One question we must ask in order to get to the bottom of the question of maximum size in any species of snake in the wild is how long do they live?

Although it's true that snakes never stop growing, this rate of growth does slow with age, so there are definite limitations depending on the age expectancy of individual species. We know that anacondas in captivity can live for 30 years, but we really don't know for sure about wild specimens.

Of course, resources available, eg. prey animals and the nutritional quality of these, will also have a big part to play in determining a definitive maximum size.

- ### Green anaconda (*Eunectes murinus*)

Snakes are surrounded by legends and mythology. In the Bible, the book of Genesis says that the serpent is the evil creature that deceived Eve into tasting the forbidden fruit and thus being the reason why man was expelled from the Garden of Eden. In mythology, however, the snake has not always been

seen as a symbol of evil. In ancient Canaan it was the symbol of the god Eshmun, the equivalent of the Greek Aesculapius - the god of healing - and connected to the underworld and reincarnation due to the snake's ability to cast off its old skin to make way for the new. I found that in Belize in Central America locals believe the Wowla, which is the local name for the common boa constrictor, is the mother of all snakes simply because it is the largest snake the peoples regularly come across, so therefore must be the mother to smaller snakes even if they belong to an entirely different variety. They also firmly believe that the Wowla is venomous after sunset. When the first Spanish explorers came to South America they named the green anaconda 'el Matatora', the bull killer. Are we really to believe that the early Spanish explorers really witnessed the spectacle of a giant anaconda constricting and consuming an adult bull, or did the name 'el Matatora' derive from over-active imaginations? It is widely known that anacondas do catch and consume capybara, wild pigs, and even on occasion

jaguars, so why couldn't a giant individual living in some rarely used, slow moving back river consume prey of bovine proportions?

Anacondas are mainly aquatic, but when this species does crawl out onto the land it struggles to pull around its massive bulk. Its circulatory system is also under great stress when crawling about terrestrially due to the snake's immense weight (this would normally be supported by the specific gravity of the water). The method used by this species to move terrestrially is called rectilinear locomotion or 'the caterpillar crawl'. The snake uses two powerful opposing muscles connected to each rib to pull its body along, and this form of locomotion usually accompanies lateral undulation. The ribs do not actually move, only the muscles beneath the skin do. The normal method used by smaller snakes is by actually walking on their ribs, which are again connected to muscles that are attached to the scutes or belly scales. However smaller snakes do also use rectilinear locomotion. The anaconda moves with incredible agility and grace when in the water, with blood circulation also being increased significantly.

For this reason alone I am prepared to believe that the real giant anacondas are likely to spend more of their time in the water than their common counterparts. They only really crawl onto land when thermo-regulating, attempting to locate another water source during the dry season, or when locating terrestrial prey items, thus spending much of their time hidden in dark murky swamps, undetected and away from human activity. If this theory is correct, then this is a positive outcome because this species is - like many animals around the world - declining in overall numbers and may be facing extinction.

A green anaconda of 30ft+ would most likely be a female as males do not usually attain these sizes.

> *"The acquisition of energy in the natural world involves a complex interaction between the biophysical environment in which an animal lives, resources available and there distribution, the social system and how it might constrain access to resources and consequently*

mating success, and the risk involved in acquiring resources."

Where better for a snake to achieve colossal sizes and successfully support their weight, than in an aquatic environment? Let us hope they remain there indefinitely.

- **Giant anaconda or *Sucuriju gigante***

As mentioned previously, zoology recognises the green anaconda to be the largest known species of snake in the world, and whilst this is likely to be correct, could there be another larger ecotype somewhere in South America's vast rainforests and backwaters? Some believe there is. What of the giant anaconda?

The giant anaconda is reported to be a colossal anaconda type Boid which may (or may not) be a new species/sub-species. It has also been claimed to be a living descendent of the prehistoric *Gigantophis garstini* from the Eocene epoch (54.8 - 33.7 million years ago). However, we must consider *Titanoboa cerrejonensis* from the Paleocene epoch (65.5 - 55.8 million years ago) as another suggestion simply because *G. garstini* was from what is today Egypt and Algeria, whereas *T. cerrejonensis* is known from Columbia. This alone makes this more likely, even though this species apparently became extinct a considerable time before *G. garstini*. My personal belief, however, is that it is far more plausible that this enormous anaconda-type snake is just that - an undiscovered variety of the green anaconda.

Maybe one day a lucky explorer will get the chance of a lifetime, and actually observe and clearly film one of these giants basking in the early sun on the water's edge. And whether at 50ft (15.2 metres) or 150ft (45.7 metres), it will be a great discovery and one that hopefully will permanently link the disciplines of natural history and cryptozoology for good.

The camoodi, a giant horned anaconda, is also reported. However, these could just be very large green anacondas that have developed extensively-wide heads with age, and the paired light and dark eye stripes that normally run back at slight angles behind the eyes have moved even closer onto the top of the snake's head. And as the jaws widen, they create the impression of horns when viewed briefly from above when travelling through water. However the camoodi is usually reported at lengths of only about 20ft (6.09 metres) which is not a considerable size for the green anaconda, so maybe it's something more interesting after all.

BRITAIN
- **Monster in the Thames**

In February 2009 Abraham W. claimed that a photo he had posted on the *Mysterious Britain* website was that of a genuine giant snake swimming in the Thames. The enormous river snake is claimed to be real and it does appear that ships in the photograph are having to swerve in order to avoid contact with it. One on-line reader stated that it resembled a pipe or mudbank more than a mystery serpent, but I think it looks far more like the work of the *Photoshop* computer programme. Another very similar photo, also released in 2009, reportedly shows another giant snake in the Amazon in South America, but this again looks to be a digitally enhanced hoax. On some websites this photograph was claimed to have been taken in Borneo.

There have been many examples of escaped snake species in Britain existing in the wild, and here is a brief list of some examples:

- In 1966 on Canterbury Road in Croydon, the RSPCA were informed that a 10ft (3.0 metres) python of unknown species was on the loose in a garage building and was surviving on rats. Firemen were called in to remove the floorboards but no snake was discovered.

- An African rock python was discovered in a pillow case in a lane in Walker, Newcastle in May 2012. A 12-year-old boy bravely took the animal to a local pet shop where it was described as very aggressive.

- A Burmese python was seen in the morning in the undergrowth at Lingswood in Northamptonshire in March 2010.

- A 12ft (3.6 metre) long python of unknown species was rescued by the fire brigade after it got stuck under a shed in a garden in Cornwall in April 2010.

- A woman with a phobia of snakes (ophidiophobia) found two living in her flat in Bournemouth. The first, a 4ft (1.2 metre) corn snake (formerly named *Elaphe gattuta* but now *Pantherophis guttatus*) was caught and taken to a local pet shop, the second similar sized Honduran milk snake (*Lampropeltis triangulum*

hondurensis) was killed with a hammer
after biting the woman's daughter twice on the hand
as she tried to catch it. Both these snakes are believed
to have escaped from a local pet shop.

WARNING:

The Honduran milk snake displays very similar red,
black, and yellow bands to the highly venomous
coral snakes of North America, and is actually the
very same as some other coral snake species
from elsewhere in the world. Unless a
100% positive identification can be made, snakes of
this appearance should not be handled by the public.
A rhyme that can help differentiate the harmless
milk snakes from highly venomous North American
corals is: Red to Black venom lack, Red to Yellow
kills a fellow.

- Police warned parents to keep children and pets
  indoors after a hungry 7.5ft (2.2 metres) boa
  constrictor called Diva escaped from her vivarium in
  her owner's home in Broom Crescent, Ipswich.

Snakes native to Britain are adapted to living in
colder conditions. Any tropical species that ends up
outdoors in Britain will need to find an area
to effectively regulate their body temperature in
order to survive (eg. move, shed skin, catch and
digest food), which is what some of the previously
noted escapees were attempting. I have been
informed recently that a large Burmese python is
surviving down a rabbit warren near the city of
Birmingham feeding on the rabbits, but I have not
yet been able to validate this. However we
can safely surmise that this is a very unlikely story,
as the Burmese python would not survive well in
these conditions as big snakes tend to need a more
continual heat source to metabolise effectively.

So far it appears some smaller individuals
are actually surviving here.

The Aesculapian rat snake (now *Zamenis
longissimus*, previously *Elaphe longissima*) has been
living well in a small area of Wales since the 1970s,
with a steady population at Colwyn Bay. It's
believed that the British ancestors of these snakes
escaped from a local zoo where they used to
be imported from Italy. There have been no serious
reports of this invasive species outside Wales.

One species put forward as a potential survivor is
the Amur ratsnake (*E. schrencki*) as they can digest
food and generally do well at temperatures of about
75°F (23.9°C). They also hibernate for 3-5 months
of the year so could over-winter underground in a
rabbit warren, where it is a fairly constant 55-57
degrees Fahrenheit (13-14 degrees Celsius) and
possibly survive and breed here indefinitely due to
the thermal inertia of the warren.

So what about large snakes surviving here?
Well this is very unlikely, even though it's true that
larger snakes would take longer to cool and,
therefore, could remain active longer than smaller
species in the British climate. It is also true that they
would also take longer to re-heat, and they would
have to find quite a considerable
heat source to successfully achieve this, such as a
man-made heating system. This is why I believe it's
more likely that big snakes are to be found in and
around, or even underneath, cities rather than in a
natural environment.

As we are all well aware, climate change
is increasing the overall yearly temperature, which
will eventually make it easier for these animals to
survive here. Currently though, it is very difficult for
most non-native snakes to survive, even ones that
could potentially tolerate relatively cool climates
such as corn snakes, as they do not tend to attain
the ambient heat available in order to successfully
digest food. Put simply, the digestive
process slows drastically as the snake cools, and the
meal rots within the snakes gut leading to a slow and
painful death.

NORTH AFRICA
- **Unknown crested snake**

Very large crested snakes are reported from eastern
Morocco to Tunisia (see also crowing crested cobra
- Central/Western Africa). Could they somehow be
surviving pythons? Or even something completely
new? The crests reported may just be where old or
unhealthy individuals are struggling to slough (shed)
their skins completely with some skin retained
around the neck and head, and building
up over successive moults until they eventually
resemble very old crested snakes. I have researched
these cryptids in Morocco and found they are still
reported.

CENTRAL/WESTERN AFRICA

- **Rock python (*Python sebae*)**

The rock python is a large species from sub-Saharan Africa. There are two sub-species, one from Central/Western Africa and the other from South Africa.

P. sebae is Africa's largest species of snake with specimens reported (but not confirmed) reaching and exceeding 20 ft (6 metres+). Although these estimations haven't been confirmed, they are considered by most to be quite possible.

This species lives in a variety of habitats near water from forest to near desert, and they have been known to eat antelope and even crocodiles, which they kill via constriction. This species has killed humans and actually eaten children.

A Uganda newspaper reported in 1951 that a 13-year-old child had been killed and consumed by this species. Although the child was eaten, he was later regurgitated. In 2002, a 10-year-old child was confirmed to have been eaten in South Africa.

Could this species, or a similar one (albeit a larger type) be responsible for the giant snakes reported in much of central/western Africa?

- **Crowing crested cobra**

Stemming from folklore similar to the basilisk or cockatrice, the crowing crested cobra is considered to be a large snake, but with three very significant differences. This mystery serpent is said to not only crow like a cockerel, but to also have a blood red crest on its head and wattles, just like a chicken. However this mystery snake doesn't bear a traditional cobra's hood. It is sometimes said to be simply a snake with a crest of feathers.

Like the fictional cockatrice, this snake is said to come from an egg laid by a chicken and hatched by a toad. This creature has a huge range, or shall we say the folklore that surrounds it states that it does; not only is it reported by natives over much of the Dark Continent and Asia, but there is also a much smaller neotropical version reported from the Caribbean, typically on Jamaica. The crowing crested cobra has also been spotted by many respectable western explorers and travellers.

A medical doctor reported seeing one of these unusual snakes in 1829 on the island of Jamaica. A snake with wattles was shot and killed in 1850, also on Jamaica.

It is reported over much of East and Central Africa and known by many names such as the bubu (Shupanga), hongo (Chi-ngindo), songo (Chi-yao) and the mbobo to the Babwe natives of Zimbabwe. Richard Freeman, the zoological director of the CFZ, recently came back from an expedition to the Garo Hills, which had been in search of the Indian Yeti or the mandeburung. Richard came back with the tail of a giant crested snake called the Sankuni that he likens to the Naga - a large crested serpent from Thailand.

The African species is reported to measure up to 20ft (6 metres) just larger than the largest king cobras (*Ophiophagus hannah)* and if discovered to be a real flesh and blood animal, would become the largest venomous species of snake in the world. The crowing crested cobra is reported to be brown or greyish-black with a scarlet head, although its new world counterpart is described as a dull yellowish brown with dark spots and only reaching a maximum of 4ft. For an extensive, detailed article on this creature, I highly recommend you reading Dr Karl Shuker's *Extraordinary Animals Revisited* (CFZ Press 2007).

- **Nguma-monene**

The nguma-monene is giant snake-like creature that has been reported near the Dongu Mataba; a tributary of the Ubangi River deep in the Republic of the Congo. The evidence exists in the form of two sightings made by Pastor Joseph Ellis, one in 1961 and the other in 1971. Ellis claimed that the visible tail part of the creature was 10 metres in length. These sightings were recorded by the late University of Chicago biologist Dr Roy P Mackal who himself led two expeditions into the Likouala swamps searching for evidence of the supposed surviving sauropodian dinosaur mokele mbembe back in the 1980s.

Similar mystery snakes are also reported in other parts of the Republic of the Congo. At Bouzoum (capital of Ouham-pende) it is known to the native peoples as mourou-ngou. Dr Mackal concluded that nguma-monene probably wasn't a snake at all, but rather some form of giant lizard that had a low slung body.

Maybe a giant species of monitor lizard?

SOUTH AFRICA
- **Inkanyamba**

In the northern forests near Pietermanitzburg in South Africa, legends are whispered of a giant serpent called the Inkanyamba. Most frequently sighted at Howick Falls this creature is reported as being about 20ft (6 metres) long, mainly aquatic and very dangerous. According to the peoples who live in the area, the Inkanyamba is a snake-like animal but with a head that looks more like a horse and is shrouded in mythology and folklore.

Inkanyamba's are believed by some researchers to be giant eels rather than snakes and there is evidence to suggest eels do reside within the falls. Howick Falls is also known for the large number of suicides that have happened there and some bodies recovered from the Falls show signs of being scavenged upon by this family of fish; could a giant unknown species of eel be responsible for the sightings? Or is the Inkanyamba purely a folkloric animal?

- **Grootslang:**

Richtersveld is home to the Grootslang (giant snake); a 40ft (12.1 metres) long 3ft wide cryptid snake-like creature that is said to reside in a local cave regularly referred to as a bottomless pit. The Grootslang is said to devour elephants that wander too close to its cave. However this account, considering the snake's reported size, is highly improbable. It is not just wandering pachyderms that seem to fall prey to the Grootslang, as in 1917 a British businessman by the name of Peter Greyson went on an expedition into the cave of the Grootslang to investigate claims that the cave was actually the opening to a 40-mile long subterranean tunnel that leads to the sea. And, interestingly, Peter Greyson's investigation into the claim that deep within the cave the walls were supposedly covered with thousands of diamonds was never proved because he was never seen alive again. Maybe he fell prey to the Grootslang!

EAST/SOUTH EAST ASIA.
- **Reticulated Python:**

Python (*Broghammerus?*) *reticulatus* is a beautiful snake with a wonderful, complex geometric

colour pattern that incorporates different colours. On the back they have a series of uneven diamond shapes flanked by smaller markings with pale centres, although over their range their marking, colour and size vary quite significantly. They feed on mammals and birds and are known to eat primates and pigs.

This species is generally believed to be the longest accepted species of snake in the world, the largest officially recorded being 32 ft (9.7 metres) in length. Reticulated pythons often reach 25ft (7.6 metres) but in 2002 a supposedly 49ft (14.9 metres) specimen was captured in Jambi Provence on Sumatra. In time, the claim was tested but not affirmed.

P.reticulatus can be a very aggressive species both in the wild and in captivity, and fatalities have indeed happened. However most reports state that although reticulated pythons have been known to kill humans and even attempted to consume them on occasion, they have rarely completed the grizzly task of swallowing the individual so regurgitated the meal. There are very few reports of this snake actually eating an adult human, although here are two reports; one a captive specimen, and the other a wild specimen of this species consuming human beings.

According to Mark Auliya, the body of 32-year-old Mangyan Lantod Gumiliu was recovered from the stomach of a 23ft (7 metres) reticulated python on Mindoro in 1998. In 1932 Frank Buck wrote in his book *West Cargo* about a teenage boy who was eaten by a pet 25ft (7.6 metre) reticulated python in the Philippines. Sadly the victim turned out to be the son of the snake's owner.

One specimen named Samantha was reported to be the largest snake in captivity. She resided in the Bronx Zoo in New York. In 1910 Teddy Roosevelt offered a reward of $50,000 to anyone who could deliver a live snake over 32ft (10 metres) in length to the zoo and so far only Samantha came close to the required size, and even she fell short of the mark. In 2001 she was measured at 25ft (7.6 metres). However up until April 15th 1963 it was believed the largest captive specimen was to be found in Highland Park Zoo (now Pittsburgh Zoo & PPG Aquarium). Named Colossus, this snake was claimed to be 27-28ft (8.5 metres) but when she died and was re-measured she was found to be smaller than previously recorded.

The largest known captive specimen (as far as I am aware) from Britain was Cassius from Knaresborough Zoo in Yorkshire. He was caught in Malaysia and taken to the zoo in 1972 and by 1978 was an impressive 27 feet (8.2 metres). This species has been reported far out to sea where they have also colonised many small islands, and are likely to be the answer to some sea serpent sightings in south east Asia.

A photograph purporting to show a 55ft (16.7 metres) "Retic" was taken in the Chinese province of Jiangxi. It was claimed to be one of two giant pythons discovered by workers clearing forest to make way for a new road near Guping city. This photo is obviously a perspective hoax as if one looks carefully you will see that the bucket on the digger that is holding up the snake is a lot closer to the would-be photographer than the main body of the machine, therefore making the snake in question seem much larger than it really was. I propose that the snake in this image is only about 10-15ft (max 4.5 metres) in length.

AUSTRALIA.
- **The Bluff Downs Giant Python:**

The Bluff Downs giant python (*Liasis dubudingala*) is an extinct genus of 32 ft (10 metres) giant python from Queensland, Australia that lived during the Pliocene Epoch (5.3-2.5 million years ago). It's believed that its nearest extant relative is the olive python *L. olivecea*. Discovered in 1992 and only known from isolated vertebra unearthed in the Bluff Downs region in north eastern Queensland, which - during the time of *L. dubudingala* - was an extensive wetlands bordered by patches of closed forest, the scientific name *Dubudingala* comes from the Aboriginal Gugu-Yalanji Duba for "ghost or spirit" and dingal "to squeeze". It's generally believed that the arrival of humans was the main cause of the extinction of Australia's megafauna with most species disappearing around 46,000 years ago. However, *L. dubudingala* went extinct around 4.5 million years ago, long before modern man arrived in Australia.

An Australian sheep farmer was puzzled at the disappearance of sheep on his farm. After a few weeks of sheep disappearing he decided to put up an electric fence to put an end to the problem and soon he had caught and killed a huge python. It turned out it was a hoax! The photograph was not taken in Australia, rather it was taken in the Silent Valley Game Ranch in Limpopo province, South Africa and shows a large rock python biting onto an electric fence having recently eaten an impala.

A large mysterious red bellied snake was killed by a farmer near Camden south of Sydney.

A tiger snake *Notechis spp.* broke the records in 1968 when a 15ft (4.5 metres) specimen was captured near Mount Victoria in the Blue Mountains. This species normally only reaches approximately 7ft (2.1 metres) so this specimen can truly be considered unusually large, and much larger than its holotype.

End note.
Even though prehistoric giant snakes such as *Titanoboa cerrejonensis* and *Gigantophis garstini* undoubtedly once existed and can truly be considered giant snakes, I firmly believe that the world's largest ever snake is not restricted to pre-history, but is alive and well and living in the dense forests of South America.

References

- Shuker K.P.N. *Extraordinary Animals Revisited* (2007) CFZ Press: Bideford.

- Up de Graff, F. W. *Head Hunters of the Amazon*. Herbert Jenkins Ltd. 1923 ("The Anaconda"). (Annotated edition, CFZ Press 2012)

- Vitt, L. J. & Caldwell, J. P.(2009) *Third Edition Herpetology* London Academic Press.

Electronic References

- African rock python found in lane in walker, Newcastle. www.bbc.co.uk (2012).

- Australian monsters. www.mysteriousaustraslia.com (1976?)

- Giant snake in electric fence photograph. www.hoax-slayer.com (2007).

- Hunt for giant snake that ate Durban boy whole. www.telegraph.co.uk (2002).

- Missing hungry Boa Constrictor found at Ipswich house. www.bbc.co.uk (2011).

- Monstrous river snake in Thames! www.mysteriousbritain.co.uk (2009).

- *Python sabae*. www.wikipedia.org (2007).

- Python seized from Croydon North house. www.maroondah-leader.whereilive.com.au (2011).

- Python spotted in undergrowth in Northampton. news.bbc.co.uk(2011).

- Python trapped under garden shed in Cornwall. news.bbc.co.uk (2011).

- Uganda newspaper report 1951. www.youtube.com (2009).

- Woman's terror as snakes escape from pet shop next door and invade her

Red Water: The man-eating lake monsters of Russia

RICHARD FREEMAN

house. www.dailymail.co.uk (2011).

Most people who have studied the history of the Loch Ness Monster will be aware of the story of Saint Columba confronting a man-eating monster in the River Ness and driving it away with the sign of the cross. The story is almost certainly apocryphal, spun to give the missionary more clout in converting Picts.

There is a story from Wales of a man-eating lake monster. Llyn-y-Gadair is a small round lake near to Snowdon. In the 18th Century a man decided to swim across it. His friends, who were waiting for him on the bank, were horrified to see a serpentine creature coiling after him as he swam. As he approached the shore the thing reared up and wound about him like a python. He was dragged back into the lake never to be seen again.

These tales are like a vicarage tea party compared to some of the stories coming out of lakes in Siberia.

Lake Chany is virtually unknown in the west but it is a vast expanse of water covering 770 square miles. It is 57 miles long by 55 miles wide but is fairly shallow at only 23ft deep with an average depth of only 6ft. Lake Chany is in the southern part of the province of Novosibirsk Oblast close to the borders of Kazakhstan in southern Siberia. It is frozen from October to April, but reaches 20°C in July.

A story has unfolded in the lake's waters in recent years that sound like the plot of a horror movie. A powerful, snake-like monster some 30 feet long has said to have killed and eaten 19 people. The story broke in the West in 2010. The attacks apparently began around 2007.

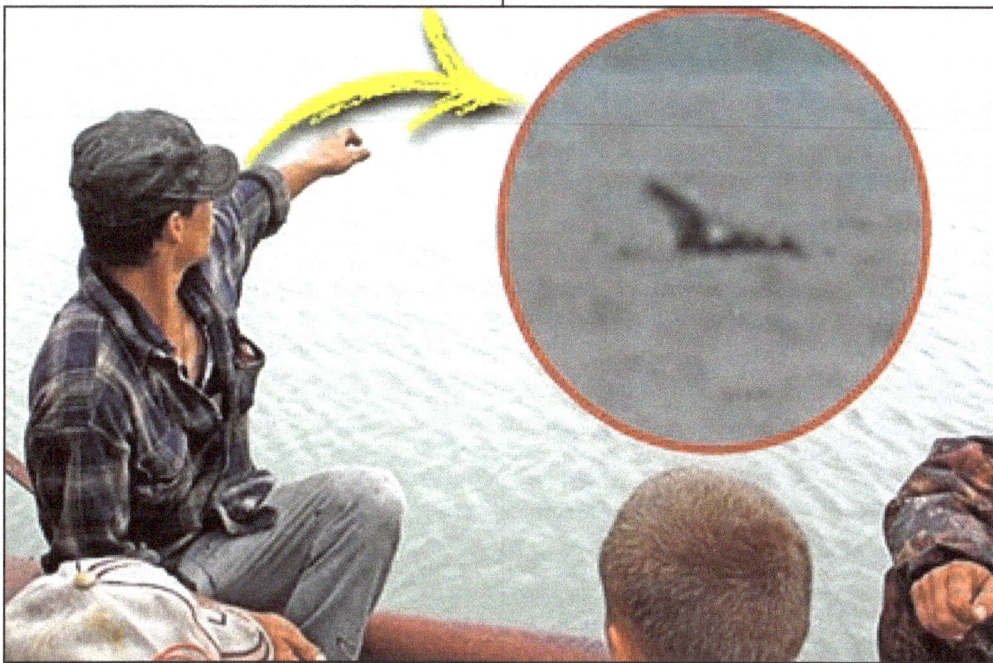

The creature involved in the attacks is described as serpentine and huge. One witness, Vladimir Golishev (60) was in the boat then the creature overturned it and dragged his friend away. He told the *Daily Mail:*

> *"I was with my friend some 300 yards from the shore. He hooked something huge on his bait and stood up to reel it in. But it pulled with such force it overturned the boat. I was in shock-I had never seen anything like it in my life. I pulled off my clothes and swam for the shore, not daring hope I would make it. He didn't make it and thy have found no remains. It's time to find out the truth."*

In 2007 a 23-year-old special services soldier, Mikhail Doronin was lost when something capsized his boat. His 80-year-old grandmother Nina was watching from the shore and said that the lake was calm. Her husband, 81-year-old Vladimir, said *"Something on an awesome scale lives in the lake, but I have never seen it."*

Official figures say that 19 people have vanished in the lake in the past three years. Locals say the figure is actually much higher and that remains have washed ashore with bite marks showing large teeth.

Fishermen demanded an official probe but the authorities passed off the deaths as 'drownings'. A blurry photograph purporting to show part of the monster was released at the same time as the main story. Like most cryptid pictures it is far from clear, but may show a fin protruding above the water.

Needless to say Lake Chany is too far north and far too cold for crocodiles. The serpentine description makes the Chany creatures sound like huge eels. Lake Chany is landlocked. It has melt water rivers running into it but none running out. Apparently eels are unknown in the lake but perch, zander, roach and pike are present. It should be noted, however, that eels often turn up in landlocked lakes and ponds and are quite capable of travelling miles overland. However, Lake Chany lies beyond its known distribution.

The community is held in terror and demanding an investigation. One wonders if an official investigation will ever take place and if it does, what it will find.

Other more remote lakes in Siberia are also said to have man-eating monsters in residence. Lake Labynkyr lies on Sorongnakh Plateau in Eastern Siberia. It is a big lake nine miles long and 800 feet deep. Despite being in one of the coldest regions on earth, the lake never freezes, maintaining a temperature of 2 degrees Celsius. Labynykr also has an evil reputation. Locals are convinced that the Devil inhabits the lake. Gun dogs that have leapt into the water to retrieve shot ducks have been eaten by the monster. One man told of how the brute pursued his raft. He described a dark grey beast with an enormous mouth. Some reindeer hunters observed the monster coil up out of the water to snatch a passing bird.

Author Gennady Borodulin also recounts a tale from Labynkyr in the 1920s in his book *In a Trip to the Cold Pole*. An Evenk family of nomads followed their reindeer and reached the shore of Lake Labynkyr. They decided to stay overnight on the shore. A five year old child went to the bank of a stream which led into the lake while adults were busy. Suddenly the adults heard the boy screaming.

> *"The father and grandfather rushed to the bank. They stopped on the edge of water and saw the child being carried away by an unknown animal to the centre of the lake. It was a dark creature, with a mouth looking like bird's beak. It held the child and moved away with quick rushes, then it dived leaving huge waves and dragged the child under the water.*
>
> *The granddad swore to revenge the 'devil'. He took a sack made of animal skin, stuffed it with reindeer fur, rags, dry grass and pine trees needles, put a smouldering piece of wood inside. He attached the sack to a huge stone on shore with a rope and then threw the sack far into the waters of the lake.*
> *At night there was noise and splashes and terrible screams of the 'devil'. In the morning the waves brought the huge dead animal, about 7m long with a huge jaw, almost one third size of the body, and relatively small legs and fins.*
>
> *The old man cut the animal's stomach, took*

out the body of his grandson, and buried him on the bank of the stream. Since then this stream is called 'The Stream of a Child'".

In 1963 a small expedition visited both of these lakes. Four members observed an object 800 metres out on Lake Labynkyr. It emerged and submerged several times. They could not take photographs as the sun was setting. The following year three teams, each replacing the other in shifts, visited the lakes. In the latter half of August, the third and final group saw the Labynkyr monster .

Two expedition members saw a row of three humps 100 metres from shore. They ran after the humps trying (unsuccessfully) to photograph them. The humps dived and rose together. It was not clear if they were separate animals or parts of one creature.

In 1964 two journalists from the Italian magazine *Epoca* visited Lake Labynkyr whilst travelling to Oymyakon. They were told that some time before, a party of men saw a reindeer swim into the lake. The deer vanished and did not resurface. Then a dog swam in and vanished as well. Suddenly, and shrouded in a mist, a vast black monster rose snorting from the lake. One of the observers, apparently a scholar, was convinced the beast was a dinosaur. The locals flatly refused to take the journalists out onto the lake.

Another story concerns a hunter's dog who swam out into the lake and was eaten by the monster. The grieving hunter constructed a raft out of reindeer skin and filled it with hot coals. He floated the smouldering raft out onto the lake. The monster snatched it and dived. It reappeared shortly after, making terrible sounds.

In the 1970s a lame horse belonging to some geologists was attacked in the night by some unknown predator. Alerted by the horse's screams the geologists got up out of their sleeping bags to investigate. They were too late. Something large and powerful had already dragged the horse down into the lake. Locals said that they often found holes in the ice with strange tracks around them.

In a letter published in *Komosomol'skaya Pravda* on January 21st 2000 Vladimir Osadchy from Moscow stated that he and a group of tourists had visited the lake in November 1979. The tourists made their way out onto the frozen lake heading for a reservoir a mile from the shore. Halfway there they stopped for a rest. An object like a black pillar was seen to rise up in the distance. A number of the tourists ran to investigate.

They reached the spot fifteen minutes later and ascended a bank two metres high. They discovered a patch of unfrozen water a metre across. The edges looked like they had been licked. It looked as if some aquatic animal had created a breathing hole. Upon returning they were told, by those who stayed behind, that they had observed the "pillar" rise several times again whilst they were gone. In the morning they searched the area and found another breathing hole with licked edges. In August 2000 a group of journalists from *Komosomol'skaya Pravda* travelled to the lake. Using sonar they detected two large moving objects at the bottom of the lake. The bigger of the two was eighteen metres (sixty feet) long.

In 2012 Associate Professor of Biogeography Lyudmila Emeliyanova led an expedition to Lake Labynkyr. She and her team recorded big objects in the lake on sonar. She told the *Siberian Times* the following…

> *"It was our fourth or fifth day at the lake when our echo sounding device registered a huge object in the water under our boat. The object was very dense, of homogeneous structure, surely not a fish nor a shoal of fish, and it was above the bottom. I was very surprised but not scared and not shocked, after all we did not see this animal, we only registered a strange object in the water. But I can clearly say - at the moment, as a scientist, I cannot offer you any explanation of what this object might be.*
>
> *I can't say we literally found and touched something unusual there but we did register with our echo sounding device several seriously big underwater objects, bigger than a fish, bigger than even a group of fish.*
>
> *Personally, I really believe that something is going on. As local residents for so many years talk about the same strange creature, it just cannot be simply invented. It means that*

there really is something. Moreover, I know the people who live around the lake pretty good and they are not liars. As icing on the cake, the ancient stories of our Nesski are much older than that of Nessie, so they can't be influenced by the Scottish sightings

There are many lakes in Yakutia and around the Indigirka River, hundreds of them, big and small, their shores are more or less populated, but all the talk is about Labynkyr and Vorota lakes, and it has gone on for many dozens of years. It makes us think about it. And these stories about the local monster are older than those about the Loch Ness monster.

As a scientist I know this is not enough to locate and study some unknown creature. I can put it like this, however. I believe there is a mystery in this lake because there is no smoke without fire.

I am sure that numerous legends which exist and circulate for many years just can't be groundless. I read many different legends but the account below is what I heard with my own ears.

Several fishermen who visit this lake from time to time say they experienced the following when fishing from a boat in this lake: during quiet, and not windy, weather when there were no disturbances in the lake, some strange waves coming from under the water suddenly heavily shook their boats.

It was as if a big body was moving under the water and producing waves which reached the surface and shook the vessel.

These stories shook me up, for instance, about a boat which was lifted by something or somebody. Two fishermen were fishing in the middle of the lake in late Autumn, they were in a 10 metre long boat when suddenly the bow began to rise as if somebody was pushing it from under the water.

It was a heavy boat, only a huge and strong animal can do such a thing. The fishermen were stuck by fear. They did not see anything, no head, no jaws. Soon the boat went down. This mysterious and very deep lake still has some secret to tell us."

In 2006 researchers using a Humminbird Piranha MAX 215 Portable fish-finder also claimed to have found something lurking in the lake. One researcher, who did not want to be named told the *Siberian Times* of his findings.

> "I switched off the 'Fish ID' and we watched just pure scanning. Soon we registered a shadow some 15-17 meters under our boat, it was about 6.5 meters long. It was pretty clear, it was not a fish and not a tree. There cannot be fish that big, and a log would have been registered in a different way. How can it swim under the water?"

In February 2013 Russian divers, led by Dmitry Shiller reached the bottom of the lake for the first time. In the same report it was stated that the team, using an underwater scanner had uncovered the jaws and skeleton of a huge animal.

Lake Vorota, mentioned above, lies twelve miles from Labynkyr. It too has a monster tradition. In July 1953 a prospecting party led by geologist VA Tverdokhelbov travelled to the Sorongnakh Plateau. His diary of the trip was published by a Soviet magazine eight years later. They visited both lakes. In Tverdokhelbov's own words…

> "30 May. We left Tomtor village, went 70 kilometres up the Kuidusun Valley, turned left and got to the large Sordonnokhskoe plateau. Ahead of us there is Lake Labynkyr where there is storage with food and equipment.
>
> There are many legends about this Lake Labynkyr. In the evenings sitting by the fire our old guide told us that a 'devil' lived in this lake. He is so big that the distance between his eyes, as Varfolomey said, 'is wider than a fisherman's raft made of ten logs'.
>
> I heard about this 'devil' before and many times. In Ust-Ner, I heard that the devil ate a dog. The dog swam to bring the shot duck to the hunter, then huge jaws raised from the water and the dog just disappeared in a moment.
>
> One of the Tomtor villagers told me that one day he found a huge bone on the shore of

Lake Labynkyr. It was like the devil's jaw - if you put it vertically, you could ride on a horse through it like under an arch. He said this jaw bone remained near the fishermen house on the shore.

> I heard legends how a whole caravan perished going under the ice of Labynkyr. It was spoken that people saw a big horn stuck out of the ice. People gathered around it on ice and tried to take it out but suddenly the ice broke and many people and reindeer died'.
>
> 5 June: Early in the morning we got to the shore of Lake Labynkyr and reached the storage. Comfortable tents with wooden beds and floor and table awaited us.
>
> **7 June:** We are having a rest. Lake Labynkyr is a square, 15 km long and 3 km wide. I found the ruined fisherman's house on the shore, carefully explored the house and all around it but did not find any 'jaw bone'.

He did not witness anything untoward in Labynkyr but went on with his expedition.

> **28 July:** Now we stopped at the shore of Lake Vorota. Mikhail made a raft and went to measure the depth. It is 60 meters as in Labynkyr. But the lake itself is much smaller.
>
> **30 July:** This is what happened today. It was sunny friendly morning, Boris Bashkatov and I went on a walking trip around Lake Vorota. We had to climb rocks on the way - about 11 am the way became dangerous and we decided to go down a bit, closer to the water. Looking at the water from the rock, I clearly saw a terrace under the water with a huge white spot on it. But when I looked at the terrace again a minute later there was no white spot there.
>
> Maybe sunshine is joking with me', I thought. But suddenly Boris shouted 'Look! What is there, in the middle?' We stopped. Some 300-400 meters away on the water there was clearly seen some white object, shining under the sunlight. 'A barrel', said Boris, 'made of tin.' 'Maybe a horse got into the lake,' I said.

Truly, the object was swimming, and fast enough. It was something alive, some animal. It was making an arch - first along the lake, then right towards us. As it was getting closer, a strange coldness like a stupor was growing inside me. Above the water there was big dark grey body, the white colour has gone. On this dark grey background there were clearly visible two symmetrical light spots looking like eyes and there was just stick in the body - maybe a fin? Or a harpoon of an unlucky fisherman?

We saw just a part of the animal but we could guess its much bigger, massive body was under the water. We could guess this looking how the monster was moving - raising from the water, it threw its body forward then fully went under the water. At this time the waves were going away from its head, waves originating under the water. 'Flapping its mouth, catching fish', I guessed.

The animal was obviously swimming towards us and the waves made by the animal reached our legs. We looked at each other and immediately began to climb up the rock. What if 'it' goes out of the water? We witnessed a predator, no doubt, one of the strongest predators in this world: such indomitable, merciless and some sensible fierceness was in every his movement, in all its looks. The animal stopped some 100 meters away from the shore. Suddenly it began to beat against the water, waves went all ways, we could not understand what was going on. Maybe it lasted just a minute and then the animal was gone, dived. It was only then when I thought about a camera. We stood for another 10-20 minutes, it was quiet. We went further.

There was no doubt, we saw the 'devil' - the legendary monster of this area. The Yakut fisherman was right, the animal had dark grey skin and the distance between its eyes was surely not less than a raft of 10 logs. But he saw it in Labynkyr and we saw it in Vorota lake. They are 20 km away from each other - and they are not connected

I recalled that white spot under the water. Obviously, the animal was hunting at that underwater terrace and we scared it when shouted going down the rocks."

Tverdokhelbov had visited Lake Vorota before in 1945 and had seen a strange animal whilst swimming.

> *"We turned around and some 30 metres away in the water we saw a huge dark grey body with two light spots and a fin above them. The animal was looking at us as if it was choosing who to start with."*

There have been suggestions that the Labynkyr and Vorota creatures are huge huso sturgeon. These giant fish can reach 24ft long and weigh 3640lbs. But these are bottom feeding fish and are not known to be aggressive towards humans. Their diet consists mainly of small fish and invertebrates. Moreover the lakes in question are outside of the known distribution of the huso sturgeon.

The wels catfish can reach 16 feet and has been known to swallow whole dogs, but once again these lakes are far beyond its known range.

Others suggested killer whales but the lakes are many miles inland and these air breathing, marine mammals would be seen breeching by anyone who spent any amount of time at the lakes.

Anatoly Sidorenkot, a Ukranian archeologist who was a member of our team on the 2008 CFZ almasty expedition in the Caucasus told us some interesting cryptozoological snippets. Some years ago a friend of his was on a boat in the Lena River in Siberia when he encountered a strange creature. It had a black humped back and a 2m tall fin. It reminded him of a killer whale, but they were thousands of miles inland at the time. A man on the boat took two shots at the beast with a rifle. It turned and swam at speed towards the boat. The man pumped three more bullets into the creature and it dived under the boat and swam away. The description recalls creatures described from Lake Vorota. The beasts here are up to ten metres long, have a dorsal fin and a wide head. Could they be some form of colossal fish?

The lake monsters of Siberia are one of the cyptozoological mysteries that I would most like to investigate. Given the resources, time and bait I

The Saga of the Bigfoot Genome

JON DOWNES

Unarguably, one of the most popular - note, we do not say *important* - news stories of recent months is the claim by a Texas veterinarian called Dr. Melba Ketchum that she and her colleagues have managed to sequence the genome of bigfoot.

Surely, we can almost *hear* you say, news of this magnitude *has* to be important? Well, yes and no. The news was, of course, *potentially* important, but because of the way that it was disseminated, and because of other information that has come to light since, this information cannot be taken at face value.

This was only one of several remarkable rumours which shot through the bigfoot community at the end of 2012 and the beginning of this year.

- There were claims that a bigfoot, codenamed 'Daisy' had been captured, and that it was languishing, and had perhaps died in captivity.
- There were at least two claims that a bigfoot had been shot; one of these involved company with whom we

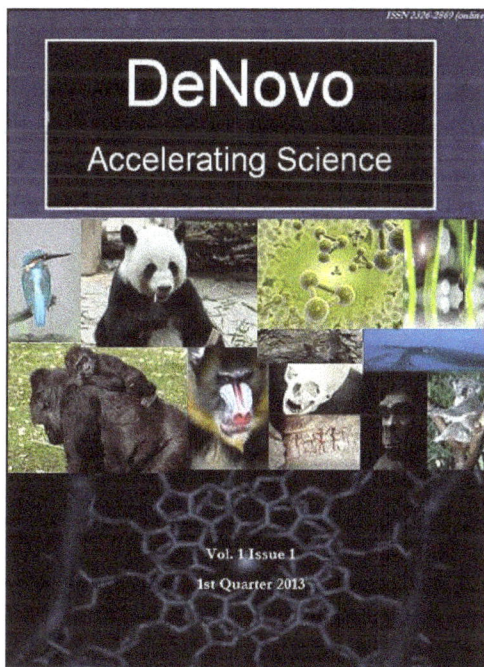

had been filming some years ago, but with whom I think it unlikely that we shall collaborate again. [1]

- The other alleged incident involved someone called Justin Smeja [2,3] who claimed to have shot a bigfoot in 2010 in California's Sierra mountains. He later

recovered some tissue samples which, upon analysis, turned out to be from a bear. [4]

After all these claims and more, excitement within the community was at fever pitch, when Melba Ketchum made her claims.

Now, let's get one thing straight. Not only do I not approve of *ad hominem* attacks, having been the subject of them too many times in my life, I believe that Melba Ketchum is probably totally sincere in her claims. I just think that she is mistaken. I also totally disapprove of the prevailing claims that because Ketchum is an attractive blonde that she cannot possibly be a scientist. These are as stupid as the claims that have been levelled at me that because I have long hair and dress scruffily that my work cannot possibly be taken seriously.

The whole story started at the end of November

1. http://www.ghosttheory.com/2012/12/14/another-dead-bigfoot
2. http://bigfootevidence.blogspot.co.uk/2012/12/statement-from-justin-smeja-regarding.html
3. http://thebigfootdiaries.blogspot.co.uk/2012/11/the-incredible-story-of-justin-smeja.html
4. http://bigfootevidence.blogspot.co.uk/2012/12/tyler-huggins-and-bart-cutino-shares.html
5. http://www.bigfootlunchclub.com/2012/11/david-paulides-releases-bigfoot-dna.html

when Igor Burtsev, a Russian yeti/bigfoot researcher announced that *"The DNA analysis of the Bigfoot/ Sasquatch specimen conducted by Dr. Melba Ketchum the head of DNA Diagnostics, Timpson, TX, USA has been over!"* [5]

According to most sources on the internet, Dr. Ketchum did not authorise this press release, and wrote on her Facebook page: *"It is unfortunate that the partial summary of our data was released in this manner, however, I will be making a formal response in the next few days. Even though Igor Burtsev released this, it was not Dr. Burtsev's fault."*

The internet rumour-mill started working overtime. Matt Moneymaker was quoted on *Cryptomundo* in early December as saying: *"Heard from a reliable source connected with an article reviewer for* Nature *(a major science journal published in the UK) that the Ketchum paper was handed back (i.e. not *rejected*) for several reasons. One of the reasons: The paper "does not contain a testable hypothesis"."* [5]

On 13th February the story broke that the paper was being published in *Journal of Advanced Multidisciplinary Exploration in Zoology*. This was all jolly good except for the fact that the journal certainly seemed to exist (despite claims on the internet) but their website is rudimentary and the links are dodgy. It seems to be published by a bunch called SCHOLASTICA [6] (who sound somewhat like a Britpop band from twenty years ago) who claim to provide a service whereby you can...

> *"Manage and publish your existing journal, or lead the Open Access movement in your field by starting a new journal. Scholastica makes it easy to collaborate on a journal and publish scholarship at the click of a button. Scholastica puts control over scholarly publishing back into the hands of scholars."*

Then a press release appeared claiming that:

> *A team of eleven scientists with expertise*

in genetics, forensics, pathology, biochemistry, and biophysics has sequenced three whole nuclear genomes from a novel, contemporary species of hominin in North America. The study, *"Novel North American Hominins, Next Generation Sequencing of Three Whole Genomes and Associated Studies,"* which analyzed DNA from a total of 111 high-quality samples submitted from across the continent, appears in the inaugural issue of Denovo: *Journal of Science* (*http://www.denovojournal.com*) on February 13. [7]

I tried to log in to the Denovo press room, but I have no press card, and apparently as I am not a journalist I am not trustworthy enough to have access. I used to do contract work for the *News of the World* so my honesty and probity should be above suspicion. So I wrote:

> To whom it may concern:
>
> Whilst I do not have an accredited press card, I am the editor of the world's oldest cryptozoological journal, and the daily online magazine http:// forteanzoology.blogspot.com/ which gets hits of between 2-5,000 every day. I also present a monthly webTV show about cryptozoology, which has produced more than 60 episodes.
>
> Would it be possible for me to have access to your press room?
>
> Yours faithfully,
> Jon Downes
> (Director of the Centre for Fortean Zoology)

I never received a reply, but in the best traditions of investigative journalism I received a leaked copy of the paper from a source which I would rather not reveal. And no-one got my zoological Woodward and Bernstein jape that it came from a source codenamed 'Deep Stoat'.

5. http://www.cryptomundo.com/bigfoot-report/mm-sasquatch-dna-project/
6. https://scholasticahq.com/journals

7. http://www.prweb.com/releases/2013/2/prweb10427105.htm

My first impression was that the whole thing was appallingly badly laid out. It looked more like an amateur parish magazine than a scientific paper, but what about the contents?

I am a naturalist and a journalist, and although I am - by definition - a zoologist, in that I am a bloke who studies animals, I never went to university, and my technical knowledge is sadly lacking. So I passed copies of the journal over to two of my colleagues whose zoological chops are far more impressive than my own.

Now, before we go any further, I am not going to identify them. Why? Because as the provenance of the said PDF is a little dodgy, and I know next to nothing about American copyright law, I do not want to implicate the two zoologists (whom I shall call Zoologist #1 and Zoologist #2) in any criminal act which I may or may not have been party to. We are NOT quoting any of the text, or posting pictures except for the front cover, so - as far as I am concerned - there should be no problems, but better safe than sorry.

ZOOLOGIST #1

I think their methodology looks weak at best, Just their description of the variation in the morphology of the hairs is enough to make me think they have been analysing hairs of several different species of animal mixed together. One of the photographs of the hairs looks distinctly human to me, whereas one of the others looks more like a bear. Apart from that the paper has a lot of statements they just expect us to take at face value, That's simply poor science. I can't see any scientific journal accepting this.

ZOOLOGIST #2

- *The English is poorly written. In a normal paper, a high standard of English is assumed, and papers can be rejected based on this. A normal reviewer would jump on this (though obviously small typos and things are found in any paper,*

as you would expect)

- *From the introduction, it is obvious that the writers are not writing from a null position. Because good science is hypothesis driven, you need a null hypothesis to act in opposition to the actual hypothesis (for instance, I hypothesise that compared to the average human, there is a significant difference between the sizes of X and Y. The null hypothesis would be that there is no significant difference). Study introductions should review the literature from an un-biased point of view, make hypotheses accordingly, then go into the study. This kinda doesn't.*

- *"Some photographic evidence also exists such as Figure 4 is a reddish brown Sasquatch sleeping in the forest" (page 2, 4th para) is one heck of a statement to give, especially without a reference to a paper (or at least a book!) demonstrating that it is in fact a 'squatch. To my knowledge, no peer-reviewed paper does so.*

- *"Video of the same Sasquatch is seen in Supplementary Movie 1 where her respirations are counted at only 6 per minute." Again, unreferenced assertion of sex and existence of the being as a Sasquatch.*

- *They end the introduction by telling us what the study finds (in their eyes). This is a poor way to write a paper introduction as it indicates the lack of a null position (again).*

- *Methods: no references for much of it. If you "thoroughly cleaned [the samples] in a manner consistent with forensic testing procedures" you need to reference the paper you got this from!*

- *When they talk about "primers", they don't show or reference the sequences used! Bad practice.*

- *No actual academics are on the paper, they are all forensic scientists. A potential bias from influence via funders and the private sector perhaps?*

- *In the results section, they make inferences and speculations about the*

data (eg, "With the wide variety of
haplotypes in the study and especially
with the majority of the haplotypes being
European or Middle Eastern in origin,
migration into North America by these
hominins may have occurred previous to
the migration across the Bering land
bridge."). Again, a stupid thing to do, that
is what the discussion is for.

- Random crappy readouts from various
  programs are strewn all over the place
  with no explanation. Again, shoddy.
- They give it a scientific name. With no
  haplotype. And the name is not in italics.
  A mistake which is guaranteed to irritate
  me.

But now the story gets truly surreal. For some
reason there are a few salient points missing on the
'press' copy of the PDF which I have, and - having
read it - I do not feel inclined to pay thirty dollars for
the privilege of having the full version. One of the
things missing is the references, which is a little
peculiar.

However, one of the references cited is apparently:

Milinkovitch, M C, Caccone, A and Amato,
G. Molecular phylogenetic analyses
indicate extensive morphological
convergence between the "yeti" and
primates. Molecular Phylogenetics and
Evolution 31:1–3. (2004)

The famous writer Peter Matthiessen notes
that "The yeti is described most often as a
hairy, reddish-brown creature with a rigid
crown that gives it a pointed-head
appearance; in size, despite the outsized
foot ...it has been likened to an adolescent
boy, hough much larger individuals have
been reported" (Matthiessen, 1979, p.
119). This is perfectly consistent with the
description given earlier by Haddock: "A
sort of enormous monkey ...with a huge
head like a coconut" (Hergé, 1960, p. 37).

Behavioural data on the yeti is also very scarce but it probably can walk upright on its hind legs and it has been recorded stealing bottles of whisky from camp sites—a behaviour that has made it called "the pithecanthropic pickpocket" (Hergé, 1960, p. 37).

The citation given is:

Hergé, 1960. Tintin in Tibet, English version. Casterman, Belgium.

The paper was a well-known April Fool's Day joke.

So where does this leave the credibility of Ketchum and her collaborators? If the fact that the paper was effectively self-published was not damning enough, the news that one of the references was a joke based around a well-known children's book, albeit one that had Bernard Heuvelmans as technical advisor, is devastating.

And according to Sharon Hill there are *two* other papers of doubtful provenance cited in the references. [8]

I am afraid that Dr Ketchum and her team have scored a massive own goal. However, I agree with Sharon who goes on to write:

To be clear, I do NOT think the paper is a hoax, nor the study. I think it showed TERRIBLE judgement and lack of professionalism. The placing of these certain papers in the document without reading them is the AUTHOR's responsibility. Excuses and passing the buck are no good at this point. Look at what she is claiming – Sasquatch DNA! You can hardly get more controversial and you don't have your ducks in a row? Unacceptable.

Unfortunately, even if the team go back and do it all again properly, their work will always be tainted with the memory of this ludicrous fiasco.

---

8. http://doubtfulnews.com/2013/02/ketchum-uses-april-fools-paper-as-reference/

The editor and his compadres welcome letters for publication on all subjects covered by this magazine. However, we would like to stress that neither this magazine, or the CFZ are responsible for opinions expressed, which are purely those of the letter writer.

A Haldon BHM?

Dear Mr Downes

I thought an incident that occurred several years ago on the edge of Dartmoor may be of some interest to you. I have never really been able to make much sense of it but thought you may be interested.

I'm writing this account around 12 years after it occurred however I can remember it quite clearly, I think it was around 1998 – 1999, I was staying at the Haldon Lodge caravan site near Kennford in Devon. I spent a great deal of time in the Haldon hills especially around the Bird of Prey watching point and the surrounding pine woods. On one occasion while alone I explored a really dense area of pine trees, literally moving on hands and knees to get through the area. I was heading for a clearing on the hill side that was visible from the caravan park. After maybe half an hour or more I came across a huge clearing surrounded by trees. It was in this area that I encountered something that I could not really work out. Shortly after emerging from the tree line I immediately spotted what looked like a large hairy shape sitting on a tree stump. At first I thought it looked like a man in a fur coat however it was a blistering hot day. What ever it was I believe I was seeing it from behind. I was a little on edge but decided to move a little closer. On doing so I noticed what looked like the back of a head although only the top, again this appeared to be hairy and connected to the main body and I noted that its colour was a brown/ grey colour. I could not tell if it was fur or hair but could now get a good gage of its size. What ever it was it was big, easily the size of a very large man and very wide. It was at this point that I took one step closer and what I presumed was the back of the head moved down and slightly to the side as if moving around. It was at this point I had had enough and moved back towards the trees what ever it was it had now changed position and appeared to be looking in my direction. Once in the trees I ran as quick as could towards the road.

As I have noted the clearing was really hard to get to, took a great deal of time to reach and surrounded by dense trees on all sides so I was at a loss as to what this was, I can't really see this as a man in a costume as this was not an area people visit and again it was very hot. I revisited the area several years latter and located the tree stump in question. It was apparently clear that what ever it was huge. When working out the distance between where I was standing and the size of the stump, it was clearly very big, bigger than an average sized male.

Do you know if there have been any other or similar sightings in the area or parts of Devon if so I would be interested to hear of them?

Kind regards

Ben

The Haldon Hills are an ancient, and rather strange place. The CFZ carried out a number of investigations between 1996 and 2005, but these were after reports of mystery big cats, and a wolverine-type creature. Over the years I collected quite a few accounts of strange occurrences on the hills, ranging from unexplained murders to UFO sightings, and from ghosts to alleged alien abductions, as well as the aforementioned strange animal sightings. Many of these appeared to have their epicentre around an

unofficial pet cemetery which flourished there for many years until it was closed down by DEFRA in the wake of the 2001 Foot and Mouth outbreak. Some of these are included in a book I wrote with Nigel Wright back in 1999, called *The Rising of the Moon* which is currently available through Xiphos Books, and another booklet I wrote at the same time for Bossiney Books called *Weird Devon*.

However, this is the first that I have ever heard of BHM-like phenomena in the area. I would be very interested to receive more sightings, and as (and when) I do, they shall be chronicled in these pages.

Influence of the Devil Birds

Hi Jon

Reading your book about Puerto Rico, [*Island of Paradise*, CFZ Press, 2007] I was intrigued by the descriptions of the giant birds with vulture-like features, teeth in their beaks, and "puppet-like" qualities. It recalls one of the oddest monsters to appear in a 1950s horror movie, the bird from "The Giant Claw". It has a well-deserved reputation for being the goofiest monster to ever appear in an American film, but what's interesting is that it was not made by an American special effects company but a Mexican one. While it does not have a horned beak,

I wonder if its appearance was influenced by big bird folklore in Latin American culture?

Robert Schneck

Caiman? Geddit?

I trust you are well and CFZ is thriving.

Last month I returned from expeditions in Bolivia and Ecuador. Whilst in a remote part of North East Bolivia, I met with an old friend, Matteo Nichols, who lives with his wife and thirteen children on the Rio Madidi. Matteo is a former American soldier, turned priest, who has set up a homestead in the jungle and savannah of this rarely visited area.

Living in the wilds, he and his sons frequently encounter the fauna of the region. They hunt and fish to help sustain the family. In May this year they encountered a strange animal which they had not met in the nine years they had lived there. Three of the sons, including Tom (20) were fishing on a lagoon (probably an ox-bow lake) near the river. It was around 10.00 hours when they saw what they thought was a large caiman swimming towards them. As it came closer they saw it had a wide green/purple head with 3inch horns rising from above the mouth and steam puffing from vents in front of the eyes. The back was green and yellow and carried scales rather like an armadillo. A tail stretched out behind and the total length of the creature was around 6 metres. As it appeared to be coming for them they fired a round of .22 from their rifle to scare it. The round hit the water and the beast blew more "steam" before turning away. It had been within about 15 metres.

All the Nichols are familiar with caiman but this animal was unlike any they had ever seen. Apparently they have some photographs of it, although when I met them these were back at their homestead. However they have promised to send me copies.

Local people to whom I have talked, felt it might have been a huge fish, known as a Paichi or a Pirarucu, that is found in the area. However this does not explain all the features they noted.

I've enclosed a copy of some sketches that they made, with my help and would be interested in your opinion of this sighting. I will let you see the photos if, and when, they arrive.

With good wishes

John Blashford-Snell

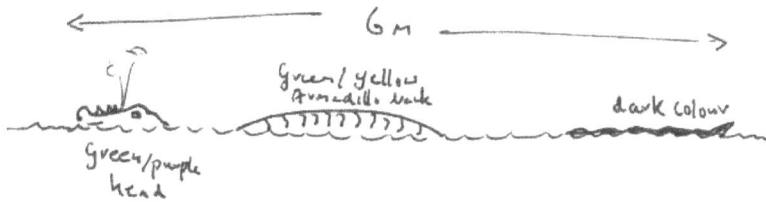

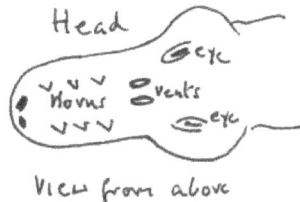

The strange creature near the Rio Madidi, Bolivia seen in May 2012

Skunk Ape Semester
Mike Robinson
 $3.99
Solstice Publishing
Website: www.skunkapesemester.com
ISBN 978-1-4657-0106-0

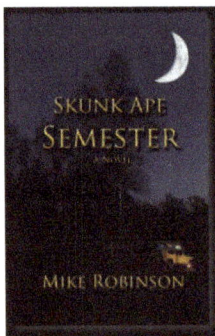

Some books take you to places you never expected to go and in directions you could have never predicted. Mike Robinson's *Skunk Ape Semester* is one such novel. The narrator, Jeremy Fishleder has a brief encounter with a skunk ape, the southern, foul smelling cousin of the Sasquatch, in his youth. In the eerily written prologue an adult Jeremy recalls a period when some*thing* was seemingly haunting the swamps and forests around his Florida home. Odd smells that come and go, rocks hurled in the night and strange feelings of being watched.

Then one night the boy comes face to face with the beast that has been prowling in the shadows just beyond the light.

Fast forward to an adult Jeremy, now a zoology professor but with an abiding fascination with cryptozoology from his run in with the monster. Dwayne, a fellow Fortean, who is retiring from active research, leaves Jeremy his van and the research notes therein. Jeremy decides to take a road trip down to join a group of investigators on the hunt for the skunk ape. He elects, perhaps unwisely, to let three students come with him. So far so good; I thought I could see where this book was going. It had the feel of one of those 1970 or early 80s Bigfoot movies like *The Legend of Boggy Creek* or *The Creature of Black Lake*. I was expecting a horror romp in which the teens and investigators are picked off one by one by swamp dwelling ape-men. I've seldom been more wrong about anything!

Skunk Ape Semester unfolds like a cross between *Fear and Loathing in Las Vegas,* Fort's *Book of the Damned,* and *Scooby Doo.* The book isn't really about skunk apes but about self-discovery as the team crosses huge swathes of the USA visiting sights of cryptozoological and Fortean interest as well as meeting up with a cast of eccentric characters. It's a sort of love song to Fortean Americana. On the long and winding road they visit sights like Lake Champlain, home of the serpentine monster 'Champ', Kelly, Hopkinsville sight of the infamous Hopkinsville Goblin encounter, Dover Massachusetts, one time stomping ground of the 'Dover Demon 'and Lee County South Carolina, haunt of the Scape Ore Swamp Lizard Man.

The people they meet and the effect of the creature sightings on them seem more important than the actual monsters themselves. The creatures remain illusive and seem to inhabit a neither world between objective and subjective. As the book draws to its end Mike pulls the table cloth from under the reader not once but twice with two totally unforeseen twists that come totally out of left field. I was expecting a horror novel but what I got was quite unlike anything I had read before. *Skunk Ape Semester* is a truly unique book. RF

Shadows in the Sky: The Haunted Airways of Britain
Neil Arnold
£12.99
The History Press
ISBN 978 0 7524 6563 0

In the hands of a lesser author this book could have been mediocre at best, but Neil Arnold is not a lesser author. If there is one thing that defines Neil's work it is the amount of unseen, obscure and new material he includes. Many books on the paranormal parrot the

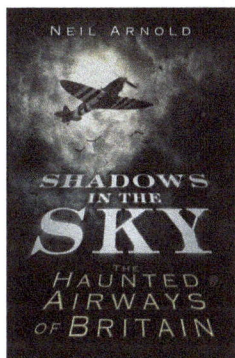

same old cases over and over again. This book could so easily have been filled with nothing more than tedious UFO reports and encounters with phantom aircraft. But what we have here is a far greater work with a far broader range of Forteana.

There is a fascinating chapter on balls of fire. Ball lighting, itself a barely understood occurrence is often used as an explanation. Though in many cases this seems to be correct, Neil rightly points out that the theory cannot explain *all* encounters with balls of light.

We also learn of phantom airships, the precursors of the modern UFO that were reported widely in America and Europe in the late 19th and early 20th centuries. Stranger still are the medieval tales of flying ships. The phenomena itself seems motile, shaping itself to the human mindset of the age.

A Fortean favourite is rains of fish and frogs. The usual sceptical explanation is that these creatures were drawn up in waterspouts to be deposited in other areas later on. This fails to account for the fact that almost invariably only one species is involved. Remarkably selective these waterspouts! Other falls have included worms, snails, blood, jellyfish, excrement, eels, pennies and pebbles.

My favourite chapter is the one that deals with winged monsters in the UK.

Herein there are accounts of that most ancient and powerful of monsters, the dragon. Sightings of dragon-like beasts have been made as late as the 1990s! As well as dragons we hear of encounters with phantom birds, gargoyle-like beasts, flying horses and the infamous Cornish Owlman.

Written in a highly engaging and entertaining style *Shadows in the Sky* is one of The History Press' very best books and makes for an absorbing read. RF

The Life Story of a Chilean Sea Blob, and other matters of importance
Theodore Carter
Queen's Ferry Press
www.queensferrypress.com
ISBN 978-0983907114

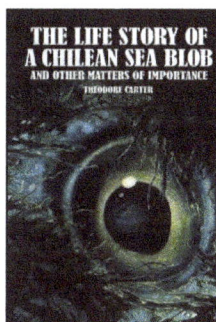

I get sent a lot of books, but seldom have I been sent one as immediately engaging as this one. This is a delightful collection of peculiar (and sometimes beguilingly macabre) short stories with Fortean themes.

Carter is a particularly elegant writer, and he provides a series of unique takes on familiar Fortean phenomena. For example in *The Who Doin' Doll* he deconstructs the familiar mythos of the voodoo poppet, and his boy who could walk on water did so for completely secular reasons.

In the same story he provides a delightful twist on modern cultural mores. The first reaction to the previously mentioned boy walking on water is that a fervent Christian journalist accuses him of mocking Christian values, and a right on Liberal complains that such actions within the public school system are mocking the separation of church and state.

In another story he looks at a young man with a particularly sensitive nose, and we discover how this turns out to be more of a curse than a blessing. And another story starts off with someone throwing up a live goldfish.

There are times, as I get older, that I begin to feel jaded, but it is the discovery of books like this, that come out of nowhere from someone that I have never actually heard of that gets rid of that feeling. Theodore Carter is a very talented writer with a spectacularly mischievous imagination, and I am very much looking forward to seeing what he comes up with next. JD

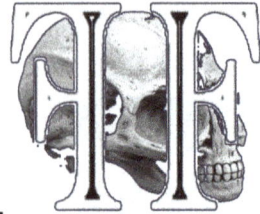

THE WORLD'S WEIRDEST PUBLISHING GROUP

W e publish a lot of books. Indeed, I think that we could quite easily claim to be the world's foremost publishers of books about Fortean Zoology and allied disciplines. However, I feel that it would be unethical to review our own titles. So here, to end this edition of *Animals & Men*, is a brief look at some of the books we have put out in the last six months.

- *The Mystery Animals of Pennsylvania* by Andrew Gable

This is a remarkable book which provides information that we had never read before. As referenced elsewhere in this issue, the section about the long-tailed mystery cats is particularly interesting. Something else particularly admirable is the way that the author examines Native American zoomythology, as well as the animal stories which came from the first European settlers, back when the area was known as New Sweden.

- *SEA SERPENT CARCASSES: Scotland - from The Stronsa Monster to Loch Ness* by Glen Vaudrey
- *Globsters* by Michael Newton

Two books on apparently closely allied subjects, which - surprisingly - overlap far less than one would have imagined. The most surprising thing is that, in an era which sees rotting cadavers like the so-called 'Montauk Monster' becoming media

causes célèbre, that it took so long for anyone to think of doing it.

- *Cats of Magic, Mythology and Mystery* by Karl P. N Shuker

We are very proud of this book. It is the first that we did in colour, and is the result of many years research from Karl Shuker, and many months hard graft from him and us putting it together. It is a remarkable book containing some of the most extraordinary pictures of felid colour morphs, and other feline curiosities that we have ever seen. Are we patting ourselves on the back? Absolutely!

- *The Grail* by Ronan Coghlan

The irrepressible Ronan Coghlan is the undoubted star of each year's Weird Weekend, but he is also a meticulous classical scholar. His dictionary of the Holy Grail collects together more information than has ever been seen in the same place before. And it has an introduction by King Arthur Pendragon!

- *Hyakumonogatari Book One* by Richard Freeman

A remarkable selection of elegantly nasty Japanese themed short stories from our very own Richard Freeman. He really is blossoming into a remarkable talent. This is also our first title available as an e-book.

www.ingramcontent.com/pod-product-compliance
Lightning Source LLC
Chambersburg PA
CBHW081231020426
42331CB00012B/3128